THE YOUNG PHYSICIAN'S GUIDE TO MONEY AND LIFE

The Financial Blueprint for the Medical Trainee

DAVE DENNISTON, CFA AND AMANDA LIU, MD

ISBN: 978-1-4834-7466-3 (sc)
ISBN: 978-1-4834-7467-0 (e)

Library of Congress Control Number: 2017913282

Lulu Publishing Services rev. date:02/19/2018

This Book is Dedicated In Memoriam to Dr. Amanda Liu

Dear Friend,

On the previous page, you may have noticed that this book is dedicated to Dr. Amanda Liu.

Unfortunately, she passed away shortly after we compiled the vast majority of this text.

With the blessing of her family, we have continued with her mission and have published this book.

We've been reflecting on her legacy since we heard of this news in late 2016. We've been absolutely shocked and devastated. Although we never met in person, we talked together on the phone a dozen times and emailed many times more than that.

Without a doubt, she was incredibly smart. What astounded us was how in the world did this incredible person pay off her student loans so quickly? She did this incredible feat that nobody else heard of- using credit cards- to pay off student loans by the START of residency.

That may sound ludicrous, but she did it by pitting banks against one another and ended up with a negative interest rate! Wow! Now, that's brilliant and amazing.

However, more important than money and knowledge I admired who she was as a person. Amanda was so passionate about her colleagues and about medicine. She loved her patients.

When we talked, she spoke with an inner fire that was stoked by seeing people healed. She so badly wanted to help heal the world. To see us all built up. For her colleagues, they saw her filled with a burning desire to help those around her and collaborate with her colleagues. She created her blog to help others climb the mountain of financial worry. We talked many times about concepts and ideas and ways we could help the physician community. She gave, gave, and gave some more. Yes, she had a heart for people.

As a blogger/podcaster, Dave was so excited to work with someone who was so eager and driven. She was incredible to collaborate with and was a bubbling fountain that eagerly embraced new knowledge and new ideas.

However, beyond all that, she was passionate about her family. Her beautiful daughter was her inspiration. The two of them created so many loving memories that Amanda blessed us with that we can still view and enjoy.

Whether they painted, did yoga, or created a financial legacy, the two of them were two beautiful peas in a pod. We look forward to the day when we can give that little girl a hug and tell her about all the amazing things her mommy did and how she impacted our lives and inspired us.

She was open and honest and transparent about her struggles, joys, temptations, and desires. She was an open book when many of us are afraid of how others might perceive us.

We are inspired by Amanda to live life fully each day. We are inspired by Amanda to dig down and look under nooks and crannies for the unexpected in finance and outside of finance.

We are inspired by Amanda to live and love with a nurturing heart. This wonderful woman left this earth and we are wondering how someone so amazing could leave us at such a tender age. She leaves us a rich legacy of love, devotion to family, and a system for prospering financially.

Over the next few pages, you will see some comments & others who have been inspired by Amanda.

First, we'll break down how to think about this book. There's a TON of info contained here.

If you want to dive right into, feel free to skip ahead and read the table of contents on page xvii.

Rest in peace, my friend.

Dave Denniston

SPECIAL THANKS

You can't raise a child without a village and you can't write a decent book for physicians without physicians. We are eternally grateful to the following physicians & thought leaders who helped us mold this book into its current form with their wonderful feedback.

We call them our 'VIP Group':

Physician On FIRE
Bo Liu, MD aka Future Proof MD
Ally Ha, MD
Grace Skemp-Dymond, MD
Andrea Macias, MD
JP Mendoza
Lauren Smith, MD
Mike Larson, MD
Dustin Jones, MD
John Roussalis, MD

HOW TO USE THIS BOOK

This book is dedicated to the next generation of physicians- medical students, residents, fellows, and newly practicing physicians.

Although, truth be told- any physician- while you are 25, 35, 45, 55, or 65 can benefit with the principles contained within this text.

As such, we have divided the book into six different sections to tackle the most common financial questions that physicians have as they transition into practice.

1. The Physician Millionaire Mindset
2. How To Become Debt Free
3. Your Insurance Guide
4. What You Need To Know About Investments as a Resident
5. The Resident's House Buying Guide
6. The Resident's Guide To Contracts

Certainly, some residents have no debt, but they want to know about investments.

Many, many more residents have TONS of debt and don't care about investments.

Some of you have life experiences that may make certain sections pretty useless- if you have a house, The Resident's House Buying Guide doesn't apply to you.

Overall, as we surveyed hundreds of physicians, these were the most common questions and concerns and thus what we covered.

We utilized tons of articles and podcasts and blog posts from our various websites. Sometimes, we wrote a chapter and then used it as a blog post or a podcast.

Dave has also published *5 Steps to Get Out of Debt for Physicians* and *the Insurance Guide for Physicians* as separate books (a previous version of 5 Steps was also included within *The Freedom Formula for Physicians*). However, we felt it was best to include them in here as well and have made additional edits and added additional content beyond what was in all those books.

So, you may have seen many of these piecemeal here and there across the web. However, you have never seen it in its entirety. To see the flow and the connection, we feel is incredibly important.

In addition, we have added all kinds of material that have never been seen in print.

Also, we've attempted to make short chapters. This was designed for you to read a little bit each day or whenever you have the time.

If you aren't a big reader, take it a chapter at a time. On the other hand, if you can binge-read, we definitely encourage you to read a whole section at a time.

Take notes. Highlight stuff. Write questions in the margins. Get interactive!

Most of all, we love hearing from our readers. Feel free to e-mail Dave by contacting him at dave@doctorfreedompodcast.com.

A Special Note from Amanda Liu's Sister

Dear Reader,

I am so glad you are reading this book. For the simple, selfish reason that my sister, Amanda, coauthored it with Dave and Jon, I would love for countless many to read and benefit from this book. What's even more selfish, I am hoping you fit a certain bill. I hope you are reading because you value learning. I hope you are reading because you're a physician (or in training) and you recognize your need for support from the financially savvy. I hope you are reading because you value your work and want to free your mind from the stress of debt in order to better serve your patients.

In fewer words, I hope you are reading because you are in many wonderful ways just like my sister was. Unfortunately, there was another side of her, dark and vast. In one of the final chapters, Dave called her "raw and real", and there really isn't a better way to describe her. In late November 2016, my sister committed suicide. All of my life before that day and after have been sheared apart. Amanda's passing taught me a lot, and I'm sure to keep learning from it.

One of my most treasured lessons from her is learning to credit intentions. I used to make the mistake of judging the (million and one) things she starts by their outcomes, and we would butt heads because I knew if she could slow down and focus, she could do amazing things, damn near perfectly. I didn't realize without the care she desperately needed, there would be no focusing or slowing down for her until the heartbreaking end.

I've since learned to take a different view to things. Now when I read this book, I see clearly her every desire to be kind, every impulse to begin something for the good of others. That's what I choose to notice now, instead of being bogged down by minute flaws.

In an email, she once wrote somewhat despondently that she felt she used to "praise more and criticize less; sing more and yell less; love more and hate less; smile more and frown less; admire more and envy less". She wanted to take the steps that would change things for the better. Even though she left us a mere seven months later, I've learned to remember that she tried.

So, dear reader, if you will read with a generous and kind heart, you will discover the true, good Amanda I love. She is nerdy. She liked medicine because it required her to keep abreast with new knowledge. She liked being frugal for the math and problem solving skills it required. She is compassionate. She cared about people, animals, and the planet. She is driven. She met and exceeded her goals in most many of the things she attempted. She is loving. The love she had for her daughter, Mini, is vivid in every line she wrote.

Amanda once wrote that "our hours are the most precious and irreplaceable asset we each have". I believe you will learn quite a bit from this book, and it's certainly worth the hours you invest in it. And if you learn anything from this preface, I hope it's the importance of self-care. If you are reading this book to boost your financial knowledge in order to better provide for your loved ones, know that investing in your own mental (and physical) well-being ensures that you can provide your loved ones with the treasure of your company longer. And check in with those around you as much as you can, or, as Amanda put it once: "spend time on things with souls, things that teach us lessons and expand our horizons."

Sincerely,

Elva Liu

What Others Have Said About Amanda

"I received a message in my inbox that was polite, assertive and smart, with the subject line, "Any tips?' Amanda had put together the DrWiseMoney website and was reaching out to me because I had done a podcast for physicians about starting your own website. As it turned out, DrWiseMoney was very well put together, and Amanda actually gave me insightful tips on how to improve my website! We spoke a few times, and I interviewed her for a 'physician success story' in which she shared a touching glimpse into her own childhood, when she collected pennies to help her family. We kept in touch, and she wrote an informative guest post for me about student loans. Amanda gave me the honor of calling me a mentor and a friend, and I truly felt that I lost a loyal, gentle and caring friend when she died so suddenly. As part of a 'group' of medical writers bound by Amanda, I have seen how she touched so many people with her genuine kindness and sincerity. She will be missed."

Heidi Moawad, MD
NonClinicalDoctors.com

"I've had the pleasure of getting to know Amanda through her persona as Dr. Wise Money. Amanda is a hard working resident physician, mother and last but not least a tireless champion for the wellbeing of her peers wearing the white coat. Her passing has left an immense hole in the physician blogosphere and my heart as well."

Future Proof MD
Blogger/Colleague at FutureProofMD.com

"Amanda was one of the first physician finance bloggers that I found online and certainly the first one I interacted with. I was just starting my blog at the time and she made it a point to reach out to me and make me feel a part of the community. I'll always remember and appreciate her tireless work ethic and warm heart."

Passive Income MD
Blogger/ Colleague at PassiveIncomeMD.com

"The news of Amanda's passing triggered immense feelings of both shock and sorrow. She and I had communicated and collaborated quite a bit this year, swapping ideas, guest posts, and encouragement. She had seemingly boundless energy and was motivated to be the best doctor, mother, and person she could be.

Amanda will be missed by so many, but her words will continue to inspire. A new post on being a small fish in a big pond was published on 11/30, and based on her prolific productivity; I believe we will continue to hear from her for months to come – a welcome gift, but a small consolation. Rest peacefully, Amanda."

Physician on FIRE
Blogger/ Colleague at PhysicianOnFIRE.com

"Amanda and I connected through our mutual interest in finance. She amazed me with her energy and financial acumen - she would try anything and it seems she pretty darn near accomplished all she tried. She was generous to praise MY knowledge any time I would answer one of her many questions even though I was in awe of her. Her curiosity was insatiable and she was inspired by almost everything. She encouraged me to start writing for PMD and shared ideas, for which I am grateful. Most of all, I admired the way she loved her daughter, and included her in so many of her plans. I will miss Amanda's big heart, her exuberance, joy, and inspiration."

Johanna Fox Turner, CPA, CFP
Fox & Co. CPAs, Inc.

"I met her in person (along with her father) when I spoke at the U of Arizona, where she was a radiology resident. She was personable, pleasant, articulate, and enthusiastic. At the time of her death, she was making a valuable contribution to her family, society, and certainly to the physician financial blogosphere. I'll miss you Amanda. May you rest in peace."

James M. Dahle, MD, FACEP, Founder of The White Coat Investor
Mentor/Blogger at WhiteCoatInvestor.com

"Amanda was a shining light to others in all that she did, whether it was tutoring undergraduates or medical students, orienting resident physicians, or educating others on finances and well being. Encouraging others to do yoga with her during lunch, humming softly as she did her work, or always looking people in the eye and smiling, she was uplifting and bright.

Amanda, this world is a different place without you. But not to worry, your legacy will live on in your daughter, your family, your friends and colleagues, those you tutored and mentored, and all those who will be further inspired by your work."

Mike Larson, MD
Colleague at the University of Arizona

I had the honor of meeting Amanda through work at the University of Arizona. We became quick friends after she generously offered to give me financial advice one day. I was immediately drawn to her contagious laughter, perpetual enthusiasm, and free-spirited nature. Her generosity was exemplified time and time again, whether it was inviting me over her house for a delicious fresh cooked meal, or lending me her compassionate ear and offering me honest advice in a difficult time. I have beautiful memories of her and her family such as going to yoga with her on a lazy Sunday morning, drawing and coloring with her daughter, and laughing when we used face-swapping technology on the phone and took pictures. I think about her and her family very often.

Her memory lives on through so many people she has touched in many unique ways. I hope to continue to honor her by focusing on the important things in life and foster my overall well-being through her initiative for improving physician mental health and wellness. I love you and miss you so much, Amanda. Your spirit lives on through all those you have inspired.

Michelle Hershman, MD
Friend and Colleague

CONTENTS

SECTION 1
THE PHYSICIAN MILLIONAIRE MINDSET

SECTION 2
HOW TO BECOME DEBT FREE

SECTION 3
YOUR INSURANCE GUIDE

ABOUT THE AUTHORS

Dave Denniston, CFA is an entrepreneur and author working with physicians of all ages and enjoys particularly focusing on residents and fellows.

His drive to help doctors came from the birth of his youngest child, Evangeline. She is his family's little miracle baby born in May 2012 four months prematurely at a weight of 12.5 oz (3.5 oz short of 1 pound!).

He has written other books on a variety of subjects that are available for sale including his last work- *The Freedom Formula for Physicians: A Prescription for First Class Financial Health for Doctors*. You can buy the Freedom Formula at a discount directly from him at _www.DoctorFreedomBook.com_.

He is planning to write at least 1 "big book" every 2 years on other subjects related to physicians. His next project is a how-to guide for entrepreneurial physicians, dentists, and chiropractors on how they can retire securely, grow their practices using reliable patient acquisition modeling, and eventually sell their practices.

He resides in Bloomington, MN with his wife of more than a decade, Cyrena, and his two children, Gabby and Evangeline.

For regular thought leadership, podcasts, and newsletters on a variety of financial subjects, check out his website and podcast at _www.DoctorFreedomPodcast.com_.

Amanda Liu, MD who blogs as Dr. Wise Money (DWM), is a second-year radiology resident (PGY3) in the Department of Medical Imaging at the University of Arizona. Amanda believes financial fitness is critical to physician well being.

As Amanda achieves her financial goals of purchasing a home (MS4), paying off her student loans (PGY1), maxing out retirement savings (PGY2), setting up a side business (PGY2), & on track to becoming retirement-eligible in 7 years (3 years after fellowship completion), she writes about and gives talks on personal finance for doctors, assisting her colleagues to achieve financial success.

Her work is featured by websites including Physician's Money Digest, Physician Financial Success Podcast, White Coat Investor and Non-Clinical Doctors. You are encouraged to join the 30-day Mindful Financial Practice with DWM @ Dr WiseMoney YouTube Channel. You can follow her at www.drwisemoney.com.

1

The Physician
Millionaire Mindset

CHAPTER 1

THE DOCTOR PRICE TAG
By Amanda Liu, MD

Why do we _rarely_ hear of heroic cardiothoracic surgeon saving the life of a patient with AAA rupture, but _instead we hear through a megaphone on social media_ about doctor scandals with a M.D. overdosing from medication or another physician swindling cancer patients out of their living days and lives?

Why do Hollywood celebrities in skimpy clothes make way more money than similarly attractive female doctors and nurse practitioners?

Why do NBA players get so many fans and makes millions than doctors with equal caliber physical and mental stamina and prowess?

Since when do patients and the rest of the society turn against their very own health-care providers?

While we unfortunately get bad press as health care providers due to a few rotten apples, which happens in every walk of life/line of profession, I want to speak up for us and reveal some facts perhaps unknown to the non-white-coats.

THE OPPORTUNITY COST OF THE PRICE TAG

We chose medicine though it clearly is not the path of least resistance.

When I look around me, I am astounded by those surrounding me in medicine, from technologists, nurses, NP's, PA's, residents, fellows, to attending physicians.

Other Callings. Any of them would be incredibly successful and wealthy if they had chosen a better paying career whether they became NBA players, Hollywood actresses, models, or business folks.

With the exception of a few nerdy-looking people like myself, there are an overwhelming amount of good looking gals and guys in the healthcare profession.

Just scroll through the Staff/resident/fellow pictures at UA/BUMC, I'd say there are few girls who would have made it big in Hollywood if they chose so.

Play Doctors versus Real Doctors. Ironically, the play doctors in Popular TV show "Scrubs" are much better compensated financially & socially than real docs and nurses.

Yet, look at these "Hollywood material girls" training and working so hard to be real doctors, working 80-100 hours weekly through their golden 20's or 30's, 4 of these years in medical school, paying 50k/year tuition for the privilege to work every day that ends in day.

Stamina of a Champion. Another example, have you ever wonder about the neurosurgeon who just finished an 18 hour surgery?

Such physical stamina could fuel success and excellence in any line of sports.

How about the work ethic of health care providers?

Anyone with the dedication to study or work 16+ hours/ day and sustain this intensity for 10-20 years would have easily made it as a CEO of any fortune 500 company.

Turning down opportunities such as NBA players, Hollywood celebrities, or fortune 500 CEO's, we strive and struggle our way to put on our white coats and to serve fellow human beings. We too are human, and we too need love and support.

We care about the well-being of another individual beyond our loved ones and ourselves. We pay our way, financially, physically, mentally, psychologically, socially, and relationally, for the privilege to serve.

THE EMOTIONAL COST OF THE PRICE TAG

Simply, we make inhuman sacrifices.

Look around us, how many marriages, relationships are broken as someone go through their medical training, from medical school to residency, well into attending physician stages.

The amount of stress from all fronts, physical fatigue, mental exhaustion, financial stress (with 300-400k of student loans snowballing at 7%), shaken confidence (from being tested/evaluated repeatedly in small and large intervals), is simply inhumane.

Now why would you treat the natural bleeding hearts of our society (those who care enough a stranger to serve them and to carry the weight of someone else's life or death on his/her shoulders) with such inhumane demands and expect nothing less than bedside manner and stellar test scores?

Something has ought to give.

Unfortunately, for those of us who chose to put another human being's well being above our own needs, we are taken for granted.

We are expected to not crumble like any other human being would under extreme pressure and run the code and bring an acutely dying patient back from the cold arms of death.

THE FINANCIAL COST OF THE PRICE TAG

Certainly, none of us are strangers to the HUGE ticket we have to pay for the entrance & then annual bills we accrue for having the privilege of being in medical school.

Then, we are forced to take menial wages in residency when you consider the 80+ hour work week most of us have.

On top of that, we are tested and charged in every way possible.

When we signed up for medicine, we signed for a lifelong privilege to be tested, evaluated, and examined into our 80's if we choose to practice until then.

From pre-SAT, MCAT, DAT, PCAT, to USMLE step I/II/III, general medicine boards, specialty medicine boards, to interval recertification/licensure exams.

There are literally tens of thousands of dollars spent on many of these levels by themselves individually with test prep and practice exams and travel and room/board.

We are the most tested profession on earth. What's worse, the people who made the laws to test us throughout our entire career are not health care professionals. This makes little sense to me. Additionally, these examinations are incredibly expensive.

FINAL THOUGHTS

As a physician, you have endured all these ups & downs. You have taken an amazing journey.

Whether you are paying the opportunity cost, the emotional cost, and/or the financial cost of the doctor price tag, let me just say that I believe in you.

You are an amazing doctor!

The cost is worth the journey, but let's work together to help the next generation of physicians.

Too many of us today say that we wouldn't go back into medicine if given the choice.

Let's consider the question…

HOW CAN WE REDUCE THE COSTS TO MAKE IT A MORE WORTHWHILE JOURNEY?

We all know about the looming physician shortage and we need to work together and collaborate to make medicine a profession that we would gladly say- I would do this all over again!

CHAPTER 2

THE DAY I STARTED WITH NOTHING
By Amanda Liu, MD

I will always remember how dreadful it seemed when I learned that the cost of attendance for my first year of medical school in 2010-2011 was about $90,000. I had never made $90,000 a year prior to starting medical school. (In fact, I still have not made $90,000 a year as a PGY2, six years after I started med school in 2010.)

I could not imagine borrowing $90,000. Worse yet, only $8,000 of that was subsidized (interest-free during medical school). The rest came with a 1-4% loan origination fee and 6.8% interest snowballing starting from day one of medical school.

Along with this dread came one of my favorite memories of my daughter who I call "Mini Wise Money".

I was crying and complaining on the phone to my mom, feeling wronged by the fact that I had to get into so much debt to become a doctor in the US. In comparison, the rest of the world pretty much made medical school free.

It's shocking how most countries around the world appreciate what aspiring doctors already give up by embracing such an all-consuming and demanding career.

She was 2 years old at the time and she tried comforting me.

"Mommy, don't be scared of monies. I am here," she said, giving me a hug. "Don't be scared of monies."

Starting With Nothing. Due to the high cost of medical school, the day I "started with nothing" was in March 2015 during my internship year when I was 30 years old.

I tried everything in my power. I worked seven jobs in college and two jobs in medical school, to "start with nothing" at the age of 30.

Meanwhile, my non-med friends already had houses, cars, fat retirement funds, college savings for their kids, and annual family vacations to exotic places.

But I'm grateful to be starting at zero at 30, rather than 40, 50, or 60..

This is How We Do It. How did I get to zero when I had $90,000 of debt and was accruing tens of thousands, if not hundreds of thousands more during medical school?

Keep reading or simply jump ahead and find out more in Chapter 11.

Anyhow, I hate debt with a vengeance.

It all started when I was about 5 years old. I became aware that some people close to me were in financial trouble. This was no small issue where I grew up. It was a big problem. In the Chinese Mafia movies

I watched, people's extremities get amputated as warnings for missed debt repayments.

The Plan. I came up with a plan. I started collecting coins around the house. Taking them out of my dad's pants pockets, jacket pockets, and finding them under the refrigerator or behind the cabinets. I was gaining momentum as my pile of coins built up. I was dreaming of the day when I would free my family from the threats of the creditors like a hero. It was my Mulan fantasy.

But… my mom caught on. She sat me down one day and said, "I appreciate your effort in collecting coins, but we need that money for food… we can't afford to save those coins, or we will have to skip dinner tonight."

The Moment of Truth. My dream was shattered. Even my most valiant and brilliant (in my 5 year old mind) efforts weren't helpful. I felt utterly powerless.

I knew not how exactly, but I knew I'd avoid debt at all costs.

My family immigrated to the US when I was 16. While I was getting ready to attend a community college, I scored high enough to get into all colleges I applied to. I applied to four: UC Berkeley, Irvine, UCLA, and UCSD.

My grandpa gave me $40,000 for college. I worked two to seven jobs depending on the semester. I graduated from UCB penniless (net worth zero) in 2007 after nine semesters and one course short of a double major.

I abandoned the idea to pursue medical school and started a family instead when I met Mini Wise Money's dad. We were not ready to be parents financially.

We got into $100,000 or so of credit card debt, which I spent the following 3 years paying off before starting medical school. This short sprint to pay off my consumer debts reaffirmed my desire to stay away from debt. I learned that debt (no matter in what country… with Mafia creditors or not) is a horrible thing.

I started medical school as a single mother with a sweet three year old daughter and $20,000 of $100 bills. My medical school cost of attendance from 2010-2014 averaged approximately $90,000 a year.

I remember vividly, crying on the phone and complaining to my mother.

I was so debt averse that I really didn't spend any money beyond the absolute necessities such as food and shelter. My splurges were for Mini's ballet lessons discounted at the community centers.

I worked 2 jobs, mostly totaling 60 hours/week on top of full time study throughout medical school. I got some government assistance for health insurance and food (which was a big deal because health insurance for Mini and me would have cost $4,000 per year otherwise.)

I also kept opening new credit cards every 12 months or so as needed to pay for expenses and ride the debt balance on 0% interest for as long as possible before balance transferring to another card (paying 3% transaction fee for 1.5 years of 0% APR, so effectively 2% annual interest rate).

As last resort, when I was not able to squeeze out more cash flow, I'd take out the minimum amount of student loans at the latest minute possible. I did this knowing that with student loans, the origination fee ranges from 1-4% and the interest rate is 6.8% the day federal government disburses the loan.

When I was getting ready to purchase a home with a doctor's loan during MS4, my net worth was -$100,000.

I had $70,000 on credit cards with 0% interest rate and $30,000 on student loans with 6.8% interest rate.

I threw every penny I had towards my student loans. This was my highest interest debt. These debts were growing and decreasing my net worth the fastest. When I have a dollar of cash, I paid my student loans.

If I get cash rewards from my credit card purchases, I deposited them in my bank to pay my student loans down further. I spared no penny when it came to that 6.8% killer interest rate.

In early 2015, I tried to refinance my student loans with DRB. I was thinking it was the best way to help me destroy my debt more rapidly since a lower interest rate guarantees that every dollar I channel towards my debt is more powerful. Thus, it would have decreased the debt principle at a faster rate.

DRB rejected me. They required that I have $3,500 a month to pay towards their bill. This was ludicrous considering all the other bills per month that I had to deal with.

Note: This happened just few months before DRB rolled out their famous resident/fellow refinancing product which Dave addresses later in Chapter 10 on page 89.

My PGY1 annual income was $50,000, and I didn't have a spouse with income. In spite of my excellent credit history and score, my unusually small student loan of 20k wasn't possible for me to refinance in a way that would work for me. The only way DRB would refinance me was with a high income cosigner.

I was pissed. I said screw this. I'll refinance myself.

I opened a Chase Slate credit card and got a $20,000 credit limit. I wrote myself a check, deposited it in my checking account, and paid off my student loans.

I like Chase Slate because it is the only card in the market that gives you 0% transaction fee for the balance transfer check within the first 60 days of account opening.

Due to these methods, I was officially free of student debt before I turned 31. All my whopping 5 figure debt is at 0% interest rate on my credit cards. This very fact would keep some people up at night. But the 0% interest makes me sound asleep...

Then, I started aggressively attacking my 0% interest credit card debts, especially when they came due. When the credit card companies are "switching" after I took the bait 12-18 months prior, I'd either balance transfer the debt with expiring 0% interest rate to a new card or I'd pay said debt off before hitting the high 16% rate.

As I realize more and more that my happiness is found in things that are mostly monetarily free, I have been just enjoying watching my net worth grow with paying down liabilities and building up assets.

FINAL THOUGHTS

At some point, you too will start with nothing. By making some smart moves and attacking your debts, you too can start with nothing.

After all, that's much much better than negative six figures in student debt.

You can do it!

CHAPTER 3

TOP FINANCIAL HABITS OF PHYSICIAN MILLIONAIRES
By Dave Denniston, CFA

Have you wondered when you think of a millionaire... what do you think that they look like?

Who are they around?

Where do they live?

What do they do with their time?

As a matter of fact, how do you become a millionaire?

What does it take to be financially free?

What do millionaires have that sets them apart?

Also, let's consider another question; are millionaire physicians different?

I (Dave) have been on this mission where I have been learning and I have been trying to grow and to understand money and how money works.

I have been reading and reading and as a matter of fact one of my mentors Dave VanHoose of Speaking Empire suggested for me to read *The Millionaire Mind* by Thomas Stanley.

It was written back at the eve of the turn of the millennium around 2000 – 2001. I highly recommend it if you enjoy this chapter!

These are the habits that I have observed in physician millionaires that *The Millionaire Mind* reinforced for me.

HABITS MILLIONAIRES DON'T HAVE

First, let's talk about what millionaires don't do.

- They don't play the lottery
- They don't take risks that aren't calculated
- They don't buy many things on a whim
- They don't take on debt unless it can help them increase their income

THE ONE FACTOR THAT SEPARATES MORE MILLIONAIRES FROM NON-MILLIONAIRES

The most critical factor that I've seen again and again is the importance of choosing a spouse.

For millionaire folks, this was one of the top reasons that they attributed to their success: having a honest, a responsible, a living, loving, capable, and supportive spouse.

When I talk about factors of selecting the marriage, the partner, time and time again we can contrast millionaires of *the general population where divorce rates are so high*…millionaires do not get divorced.

Obviously, you cut your household in half then, right?

What's fascinating is more so than anything else….

Satisfaction with your partner's financial contribution is strongly related to how you feel about the relationship for those that are non-millionaires compared for those that are millionaires.

What the millionaires said was instead that having just enough to live on was good for them. **That finances no longer came as a sticking point between millionaire couples.**

The millionaire physician can rest easy when they know that even if their spouse loses their job… even if they change careers… that they stick together!

As a matter of fact, the divorce rates for non-millionaires are so much higher. Literally twice as high as millionaires.

We have to keep our spouses. We have to keep a loving relationship. We have to find forgiveness and the guts to stick it out in incredibly difficult circumstances.

I would like to know from you…

Where do you see successful marriages?

<u>What does that look like?</u>

Do you see millionaires having successful marriages?

THE BEST HABITS OF PHYSICIAN MILLIONAIRES

The top habits with millionaires are the following…

HOLD ON A SEC…

<u>As a matter of fact, I would like you to write this down…</u>

Grab a pen so that you can mark this up…

Don't worry, I'll wait…

………………..

………………..

……………..

…………..

Alright, here we go, the top habits:

- They test price sensitivity (they negotiate bargains and discounts)
- Having furniture refurbished rather than buying new.
- Raising the thermostat during the summer (to avoid high cooling bills)
- Switching cell phones/long distance telephone companies for more economical plans.
- Never buying from telephone solicitations.
- Having shoes re-soled or otherwise repaired.

- Buying quality (to avoid frequent replacement and repairs) and holding onto it for long periods of time. (i.e. cars, furniture, etc)
- Using discount coupons when shopping
- Buying household supplies in bulk.

PRO VS. DO-IT-YOURSELF

What's interesting to me is this idea of pro versus do-it-yourself.

I've heard it cited in many places that if you are great at something... you do it absolutely all by yourself.

You save yourself the money there.

But....

If you are an amateur at something when you are a millionaire (or want to be one)- you don't do it!

Take an example of plumbing. Obviously, you can save yourself some money if you do it yourself, but...

You might cause yourself a whole host of issues.

Many millionaires are frugal, but many think that do-it-yourself concept defines frugality.

THIS IS NOT TRUE FOR MILLIONAIRES!

In fact, the real frugality is in doing some of the little things that I wrote earlier. For example, clipping coupons or refurnishing your furniture or resoling your shoes rather than buying new things every year.

This is what makes economically productive households.

MORE DETAILS ON ECONOMICALLY PRODUCTIVE HOUSEHOLDS

This is really the struggle. I want you to be frugal. It's a weird place to draw the line. Those that are frugal are walking the line between doing-it-yourself versus spending wisely.

Some of these things are really interesting in terms of being frugal. Consider that most millionaires (57% to be exact) <u>raise the thermostat during the summer.</u>

They are paying less in the heating bill, but by the same token… <u>they are not afraid to spend money, when it might save them money over time. For example, they talked about the resoling of shoes.

Millionaires are not afraid to buy a $200 to $250 pair of shoes that will last them for a lifetime and instead just get them resoled. Compare this instead to buying a $70 or $80 dollar pair of shoes every few years.

The millionaire tends to buy quality and hang on to it. They don't replace it very often.

Another really interesting thing that I see a lot of physicians struggling with is the actions taken in purchasing a home. We'll address this later on in the book.

The bottom line here is that millionaires tend to ask for discounts on commissions and various expenses. Also, consider that most millionaires do not build their homes from scratch.

They tend to buy older homes in established areas that are built well.

FINAL THOUGHTS

Thinking about this chapter and writing down the habits have definitely brought up a few aspects for me that I need to personally work on.

I talked with my wife about the things we could do better as we look at millionaire households.

I would like to know from you....

Did you learn something today?

<u>Could you imagine yourself being a millionaire?</u>

What could you improve on?

RESOURCES MENTIONED IN THIS CHAPTER

Speaking Empire/ Dave VanHoose
www.speakingempire.com

The Millionaire Mind by Dr. Thomas Stanley
www.doctorfreedompodcast.com/millionaire

CHAPTER 4

MY FINANCIAL INDEPENDENCE BY 38
By Amanda Liu, MD

S ay what? Retire by 38?

Yes, that's right. I (Amanda) could retire by the time I am 38.

Why? I've paid off my medical school debts within my first year in residency. Yup, you heard it. I'm medical school debt free! I had little parent support. I did it with blood, sweat, & tears. I live lean & mean. I keep a very close eye on the pennies and dollars.

I could retire… but I don't want to.

Dave just went over the physician millionaire habits- many of which I follow that has led me up to this point.

Let's first talk about why I won't retire at 38, even though I can.

REASON# 1: WORKED TOO HARD TO GET HERE

At 38 years old, that's less than three years out of fellowship. It would be the beginning of my professional prime as a radiologist. I don't want to retire because I'd like to use my skills and knowledge from 26 years of schooling to serve others.

I get a wonderful sense of fulfillment from draining a patient's pelvic abscess today; I will also relish saving someone's life by making an imaging diagnosis in 7 years.

REASON# 2: SELFISH REASONS TO BE SELFLESS

I want to care for others beyond myself and my daughter (I call her Mini-Wise Money in my blog). These people include my parents, the kiddo's paternal grandparents, some of my extended family, & my sponsor child Mariela.

I want to provide for my daughter more than I have been blessed with growing up. While I worked 7 jobs in college trying to send meager money home to help my parents with their 30% interest rate credit card debts, I'd like Mini to have at least 1 but no more than 2 jobs while going to school full time.

REASON# 3: LOVE FOR MY BLOG

I'd like to keep blogging at DrWiseMoney.com. Blogging costs money and a lot of time. I plan to support my blog because helping others succeed financially & personally is rewarding to me.

REASON# 4: GIVE MORE

I'd like to give more. I sponsor one child now. I'd like to sponsor more. I may even adopt a child one day, so that Mini is not so lonesome when I'm in heaven one day.

3 WAYS TO LEARN FROM ME

3 things to help you achieve financial independence/ability to retire sooner includes:

> ➤ Lower the cost of your debt by refinancing.
> ➤ Mindful financial practice.
> ➤ Make your money work for you.

Check out the end of this chapter for blog posts on all three of these subjects.

LET'S SEE HOW I CAN RETIRE AT 38

I'm currently a second year resident. There's quite a bit of time left before I finish my residency and then my fellowship. However, we're off to a great start!

I first started my calculations with my current post tax annual income of $52,671 and annual savings of rate of 56% ($29,300 total).

Yup, we're making it on less than $2,000/month.

Anyhow, I have gotten a head start on savings- my portfolio is about $32,000.

I made several assumptions in my calculation.

> ➤ My current annual expenses equal my annual expenses in retirement
> ➤ I will never draw down the principal because I will live off of my earnings
> ➤ My annual return on investment is after taxes and inflation

More than anything else, it's my current savings ability which gives me financial independence (ability to retire) in 13 years.

Savings include $23.5k in Roth IRA & Roth 403b, $1k in taxable investment, & $4.8k in home equity. Note that my home equity does not exactly grow at 5% ROI (return of investment), but I have $5k of additional savings that will make the $4.8k in home equity a bonus even if it's getting negative ROI (home value drops rather than grows.)

You can imagine that in 5 years that I would have already accumulated another $150,000 to my portfolio and net worth (not even including home equity). That also does not take into consideration any earnings. If I am making 7% a year, my portfolio could easily be pushing well past $250,000 before even transitioning to practice.

ACCOUNTING FOR THE TRANSITION INTO PRACTICE

My calculations with the jump in salary when finishing fellowship here:

My attending post tax annual income $160,857 (based on an ultra-conservative estimate of $250k annual income, when most radiologists I know are getting $300k to start at academic/VA jobs) and annual savings $137,486.

My annual savings rate is 86%, which gives me financial independence (ability to retire) in 2.8 years of finishing fellowship (8.8 years from finishing medical school.)

Living, Earning, & Saving like I do today with attending income in 4 years; I can retire in 6.8 years starting today.

Assumptions:

➢ My employee job will pay $250k/year pre-tax. (I presume a contractor job will actually allow me to save more as business deductions are numerous and liberal.)

➢ I will continue to live like I do now after becoming an attending. Like I said before, I'm blessed to be born a materially low-maintenance girl. I get free haircuts from Mini; free clothes from friends and families who can't stand my monotonous wardrobe (7 pairs of scrubs) and need more space in their own.

➢ With the same expenses, I'd save the rest of my post-tax dollars. This makes an annual savings of $137,486.

➢ 3 years of savings and even more interest & earnings at 7% pushes my portfolio to be over $700,000.

➢ If I can earn 7% a year on that nest egg, that equals a healthy earnings of $49,000 that should be sufficient to meet my needs at $35,000 a year in future dollars given inflation of 3%.

FINAL THOUGHTS

I find it extremely liberating to learn that at 38, just 7 years from now, by year 2023, I will work because I love to, not because I have to.

Financial independence does not mean that I will drop all that I love and care for today to travel the world or become a full time gardener. It simply signifies that my motive for working & not retiring will be more profound than monetary necessity.

Final fun morsel to chew on: If I live as I do now (PGY2, with 22 years of schooling) but I choose plumbing instead of doctoring, I would have reached FI (financial independence) at 31. I would have also had more personal, family, and party time even before retiring at the tender age of 31.

Point is: Financial independence is not an end in itself; it is a means to fully enjoy one's life, professionally and personally, more than otherwise.

- How many years before you reach financial independence?

- I will continue to live, work, love the way I do today when I reach FI. Would you do anything different?

- Are there things you'd like to change today that may impact your time or path to FI?

Share your insights, experiences, and questions at DrWiseMoney.com!

RESOURCES MENTIONED IN THIS CHAPTER

Mindful Financial Practice
http://www.hcplive.com/physicians-money-digest/contributor/dr-wise-money/2016/06/cash-is-king-credit-is-queen

Make Your Money Work For You
http://drwisemoney.com/2016/05/06/how-i-contributed-23-5k-to-roth-iraroth-403b-in-2015-as-pgy2/

CHAPTER 5

HOW THIS PHYSICIAN COULD RETIRE BY 45 (CASE STUDY)

By Dave Denniston, CFA

As I write this chapter (December 2016), I (Dave) am about a month removed from receiving horrible news.

My friend, co-author, financial blogger, aspiring radiologist, and loving mom Dr. Amanda Liu passed away.

At first, I couldn't believe it. Sadly, it is true that Amanda isn't with us any longer, but her words and spirit are.

Frankly, I've had a hard time picking this project back up. I've been wondering how best to honor Amanda.

In the previous chapter, I was so inspired by that section (which she originally published as a blog post) that I reached out to her and we recorded a podcast interview (www.doctorfreedompodcast.com/amanda).

In memory to Amanda who was dedicated to financial independence and mental health for her physician brothers and sisters, I've been studying a few physicians who are on track for an early retirement and doing so in a way that they have the wealth and even more importantly the health & enjoyment of life while they are doing it.

WHO COULD POSSIBLY RETIRE THIS YOUNG?

In this chapter, I am sharing a case study of a physician that can and will retire before 45.

The first person I thought of is my frequent email correspondent and fellow northerner PhysicianOnFire.

He has an incredible blog dedicate to FIRE- Financial Independence Retire Early (www.physicianonfire.com).

I read at least one of his blog posts every couple of weeks and share them freely on social media.

While he's never revealed the Clark Kent behind his Superman identity nor the Physician behind the Fire, we can learn a ton from him.

I think he's graced us with an incredible level of transparency and leaving us with a few good giggles and belly laughs along the way.

Here are five specific lessons that I've learned from PhysicianOnFire…

5 LESSONS ON RETIRING EARLY FROM PHYSICIAN ON FIRE

LESSON# 1: THE WAY YOU ARE RAISED FINANCIALLY MATTERS

In this post (https://www.physicianonfire.com/my-path-to-financial-independence/), PhysicianOnFire emphasized that his path to independence was paved initially by his parents. They taught him the value of compounding and the value of working.

He said, *"My Dad taught me the Rule of 72 when I was a kid. When I got a job in high school, they helped me open an IRA, and helped me fund it when the $4.25 an hour I earned at the grocery store wasn't enough. They also covered the taxes for a Roth conversion (but not this Roth conversion) when I was in a low tax bracket."*

I think instilling the value of money and then combining that with working at an early age and then combining that with investing set him on an early track to success.

I'm not sure what his vices were... but I know I saved up the dough I earned as a kiddo to buy a super Nintendo.

I think I probably would have been served just as well by learning that same lessons and having my parents show me what my money could be instead of saving it & then spending it.

Anyhow, on top of these lessons, PhysicianOnFire was blessed with having a history of family who made smart choices about money. He's alluded a few times that his grandparents helped some with college. Being raised by financially smart people helps in multiple ways and some of us aren't blessed with this.

However… it's never too early to start! Also, consider, what lessons are we passing onto our kids about money?

Can we break whatever generational curses may have existed in the past and instead start generational blessings?

LESSON# 2: THERE'S A TREMENDOUS RETIREMENT COST TO CHOOSING THE WRONG SCHOOL

PhysicianOnFire was not only smart in making investing decisions as a pimple-faced, gangly teenager (I resembled that remark), he acted incredibly wise in the major financial decision of where to attend school.

Luckily, he had killer grades and aced math. He had the pick of the litter when it came to undergrad schools.

While he flirted with the idea of being near fast food joints on campus, he decided to stick to an in state public entity that paid ENTIRELY for his tuition.

He said this in his post on financial independence, *"I graduated in 4 years, enjoying it so much I decided to stay for 4 more, finally leaving in 2002 with an M.D. I consider my choice of college and medical school to be an important first step towards FI.*

*Between the scholarships, the in-state public school tuition, and a college fund set up by my grandparents, I was able to finish undergrad with money in the bank. I took out loans during medical school, lived in shabby apartments next to campus, and was able to graduate with a **hefty five-figure debt**. If I hadn't had my grandparents' help, or had gone to private school at any point, my debt would have easily been six-figures."*

As most of my readers and listeners are well beyond this age, it may seem a shade too late for this advice. However, I really want to encourage you to think about it this way- what was the cost on his parents for these decisions?

Probably nothing, my friends! If anything, it's helped them to retire much, much, much earlier.

What are you willing to sacrifice to send your kids to undergrad?

Consider this: if simply ONE of your children go to a private university tomorrow and you have to pay entirely out of pocket without them being responsible for a single cent, you could easily be out $250,000 of your retirement funds versus let's say $80,000 for a public institution.

If your living expenses are $90,000 a year, that's at least two additional years you will have to work to allow them to do this.

If you have three kids, that's six additional years you'll have to work.

That's not even including the growth that you could have had in your assets.

LESSON# 3: STAY LASER-LIKE FOCUSED ON YOUR FINANCIAL GOALS FROM THE START OF RESIDENCY

PhysicianOnFire was laser-like focused on achieving his financial goals from the very start of residency.

He was determined to build his nest egg and bought a small, comfortable condo that didn't tie him down financially.

He said, "I was able to save enough during my internship for a 10% down payment on a one-bedroom condo in residency. I became a homeowner, had a nice place to live, and the place appreciated in value."

LESSON# 4: YOU HAVE TO MAKE
SACRIFICES TO RETIRE YOUNG

In this post on his story (https://www.physicianonfire.com/my-story/), he certainly worked his tail off.

He said this, *"Fast-forward to my first big paycheck. While the other 20-some new residency grads in my class spent the first week of July studying for the written board exam, I took a one-week locums position, taking call on the 4th of July. I used that check to pay off a loan I had taken during residency.*

Time was money, and I turned my time into money every chance I got. I had a 3-week job lined up right after the Saturday exam, then drove over a long weekend to a 25-weeker up north. don't think I took a full week off until late spring."

While he didn't state it outright on the post, this inevitably meant that some other area of life got neglected.

Rather than enjoying the family BBQ and feasting on ribs & hot dogs on the 4th of July, he chose to work to get his debts paid off and increase the size of his nest egg.

This meant additional sacrifices beyond the usual ones in residency. He probably missed many more birthdays and family celebrations than your typical attending physician or resident. I'm sure this had a cost in some relationships that he didn't have time to invest relational capital.

However, today he is sitting pretty and doesn't seem to have any regrets about that season of his life.

Work early and work hard so that you can have flexibility later in life.

Can you pull a couple shifts here and there AND have the discipline to use that money to fund your financial goals?

LESSON# 5: SELF-DISCIPLINE & SELF-MASTERY ARE THEY KEYS TO UNLOCK AN EARLY RETIREMENT

More than any other category, I've found in working with over 150 clients that are physicians and regular folks, living expenses are the biggest determinant of when you can retire.

PhysicianOnFire has the self-discipline and determination to keep on top of what those are and understands the impact that they can have on his retirement. This doesn't mean that he hasn't made a mistake.

As a matter of a fact, they made a six figure choice that was tremendously painful. He wrote,

"In the fall of 2015, we sold the big waterfront house, for over $200,000 less than we had into it. Yeah, that stung. But ripping off that humongous band-aid made us debt-free and more importantly, <u>financially independent</u>.

We once again have a waterfront home on the bluffs overlooking the river. We spent a lot less on this home but it suits us very well. Having become somewhat debt averse and already paying 2 mortgages at the time we moved here, we decided to sell some funds from the taxable account and buy the home with cash, <u>keeping my goal of being debt-free at 40 a reality</u>."

I took the liberty of underlining their goal and how they made a big sacrifice to keep this dream a reality. PhysicianOnFire and his wife were willing to do whatever it took to hit that goal- even if it was extremely painful.

They had mastered themselves and were willing to look in the mirror and do a self-assessment. This meant that take a loss, an extremely painful loss because it allowed them to improve their financial standing and allowed them to move forward towards their ultimate dream.

Check out what he had to say about mastering frugality here: https://www.physicianonfire.com/the-frugal-physician-self-serving-or-self-denial/

BONUS LESSON: WHO YOU MARRY MATTERS

PhysicianOnFire is one smart dude. However, it sounds like the smartest decision was who he married. He claims she is more frugal than he is!

Beyond who you marry, even more importantly is the commitment staying together. Divorce costs so much.

In my podcast 5 Keys to Unlock Happiness In Physician Marriages with Julia Sotile Orlando (www.doctorfreedompodcast.com/sotile) , she gives some fantastic advice for physicians to keep your marriage flame alight.

"Planning To Retire? Before You Do, Find Your Hidden Passion. Do the Thing You Have Always Wanted To Do." Catherine Pulsifer

HOW HE CAN AFFORD TO RETIRE EARLY

Now that we understand his mindset and some of the ways he got started on the road to success, let's break down how the numbers work and what's within his spending patterns.

Check Out Multi-Millionaire Family Tracks Spending For a Year: They Spent How Much!?!?: https://www.physicianonfire.

com/multimillionaire-family-tracks-spending-for-a-year-they-sp
ent-how-much/

PhysicianOnFire tracked his spending for a whole year (and I assume regularly) on Mint.com. Between their houses and vacations and
eating out and the stuff happens in life, they spent $74,000 a year. That's
a bit over $6,000 per month.

Make sure to check out the link above for how that broke down.

Here are some awesome habits I wrote down that they have which
has led to their wealth accumulation and relatively low living expenses:

- Cheap cell phone plan (I know my family could do better here!)
- Renegotiated their cable/dish bill (Again, I know my family
  could do better here!)
- Good public schools (no fat bills for private primary school)
- Low property taxes (lives in a low cost area)
- Bought cheap technology ($400 laptop vs. $2,000 laptop)
- Did dog swapping with friends & family rather than using a
  kennel while on vacation
- Cuts own hair
- Shop at the Goodwill and other second hand stores
- Is Self-Insured (doesn't need life insurance or disability insur-
  ance due to wealth accumulated)
- Has NO debt
- They track their spending

Here were some things that I noticed that were NOT included in
his analysis (although, he acknowledged most of these):

- Did not include health insurance
- Did not include tithing or substantial giving to a faith
  organization

37

- Did not include income or payroll taxes

Let's say that health insurance is $1,000/month if they had to get it on their own. It could easily be $1,500/month or more, but I'm assuming that they find a great deal.

I'm also going to leaving substantial giving off the table- although for my family & I, that's something we budget for and maybe you do too.

Don't give me wrong. PhysicianOnFire is a dedicated giver. He even established a donor-advised fund with over $100,000 to help right the wrongs in the world. Make sure to read his post on that!

ARE YOU READY TO NERD OUT?

Let's crunch some numbers to see how this could work!

That makes total living expenses of $86,000/year. I don't know what his asset mix is between pre-tax, Roth, and non-qualified, but I'm going to guess if like most doctors that it is heavily pre-tax with some non-qual and some Roth and that he'll do a mix of all of them.

I don't know how much money he has- but he's alluded to being a multi-millionaire. Let's say his portfolio is a grand total of $2,000,000. On a current portfolio of $2,000,000, that's a draw rate of 4.3% NOT including taxes.

Keep in mind we want to account for inflation as well. I usually target 3% for inflation. That's a total draw rate of 7.3% based off of today's values. Not too bad- I think you could make a case for retirement- but there's not a whole lot of wiggle room if you are earning 5% with a more conservative, income based assets as opposed to a more aggressive asset mix that could kill you in a downturn where you need your portfolio to live on.

However, he's still working for a while. Let's say that there are more contributions to the tune of $400,000 or $80,000 a year and he gets some growth. (I suspect he is saving closer to $200,000 a year due to his specialty, but I will low ball it here).

If the growth on the existing assets is a mere 5% a year, using simple interest rather than compounding interest of initial principal of $2,000,000 that's $100,000 per year or $500,000 over 5 years.

That's a grand total of $2,900,000.

<u>I have fancy computer programs that could calculate all this stuff and run Monte Carlo simulations- but let's keep this relatively easy to understand.</u>

Let's say that he pulls $12,000 a year from Roth and non-qual and $74,000 a year from pre-tax money. He'll probably owe at least 15% in federal income taxes and maybe another 5% in state income taxes on the pre-tax money.

That's $14,800 in taxes per year in retirement to start out.

That makes for total expenses of $100,800 ($74k basic living expenses + $12k health insurance + $14.8k in taxes). Let's call it $100,000 to keep it in simple math.

Now, we're looking at $100,000 expenses over $2,900,000 in assets. That's an even lower draw rate of 3.44%.

Let's make sure we add inflation to that of 3%. This gives us 6.44% inflation adjusted draw rate.

If he earns a mere 5 to 6% rate of return, he will barely even dip into the principal!

He has way more than enough room to account for buying a new 'used' car or to replace his roof or virtually any other obstacle that may come his way.

Plus, if college is more expensive for the kids than they thought or he finds they want to travel a little more or they want to start donating money to causes that are important to their family, he could always pick up a locums shift every now & then.

Of course, his blog is going to continue to blow up and he'll be making a ton of money from advertising- so why bother?

=-)

FINAL THOUGHTS

My friends, I'm not saying that retiring early is for everyone. As a matter of fact, you'll find out that it's not for me and maybe it's not for you.

However, regardless of whether you are 25, 35, 45, 55, or 65, learn what you can from PhysicianOnFire.

Keep up with him on his blog and learn from his knowledge.

What lessons have you learned from this case study? What are you going to do to move yourself closer to retirement?

CHAPTER 6

5 GREAT MONEY HABITS YOU MAY ALREADY HAVE TO BECOME FINANCIALLY INDEPENDENT!

By Amanda Liu, MD

In my adventures with money and medicine, I find that it is our habits that define our future.

It's the little steps that add more and more and more.

You may be sitting here reading this book and think, "How can I possibly have the time to manage my financial life and my family life and my work life??"

I have good news for you, my friends. You already have developed habits in medicine that will help you become financially independent.

Let me explain more.

HABIT# 1: YOU VALUE EDUCATION

Have you ever heard that the best pay one can get is $0/hour?

Since our time, due its limited nature, is worth an infinite amount of money, why would making $0/hour be the best pay?

It's because when we use our time to learn, our minds expand in amazing and unpredictable ways (especially something as complex, intricate, and eye-opening as medicine).

Our mind indeed is our greatest asset and will for all our lives-- regardless of how old, frail, or even disabled physically we may be— make us money if we allow it to.

Kudos to you for paying $50k a year just so that you have the privilege to work your butt off for 80 hours/week. You work and learn every day. You even fill in on short-staff holidays.

The fact that you are willing to go into $300k debt @ 7% interest rate does mean that you've got your priorities straight!

HABIT# 2: YOU VALUE HUMAN BEINGS

Each person's greatest asset is his or her individual mind. The greatest asset of our society is our collective minds.

Since you value other human beings and your relationship with them, your innate focus in life is much more internal and meaningful than most of the external stuff that costs lots of money.

Let's say by some miracle that you had some extra time on your hands. If you are like most physicians I know, your tendency is probably to take a hike with a friend rather than window shopping at the mall. If I guessed right, you are not inclined to go spend money instead you love spending your time with people.

Learning from, laughing with, and supporting those around you is awesome and mostly free. You can lead a rich life figuratively (with good friends and company) and literally (with maxed out retirement accounts, college funds, etc.)

HABIT# 3: YOU VALUE CHARITY

You love giving. You share your time, attention, care, nurturing, your most precious resource, your mind freely with your patients and colleagues.

Those who give generously usually end up with plenty. Givers are usually planners. You are able to give because you have made room for and prioritize giving.

HABIT# 4: YOU ARE DISCIPLINED & DILIGENT

If you are not, you'd have been one of those self-declared pre-med who became pre-something-else two weeks into general chemistry in college.

If you apply the discipline and diligence you master in academic and professional life to money matters, you will be a millionaire before you know or even plan it.

It is amazing that becoming wealthy is set after set of simple habits. It just requires the same discipline and diligence that you've always had.

HABIT# 5: YOU HAVE AMAZING WORK ETHIC OR YOU ARE A GENIUS

Both are helpful to making you rich… goes without saying.

As previously mentioned, we work our tails off.

This is no small feat. Shift after shift after shift. You put your nose down to the grindstone and work, work, and work some more.

You have an amazing work ethic! Take a small piece of that and apply it to your finances.

FINAL THOUGHTS

Now take these 5 wonderful habits/traits into your money matters.

Keep the good & get rid of the bad, you'll be prosperous beyond your wildest imagination.

You can do it!

CHAPTER 7

HOMER SIMPSON'S GUIDE TO DEBT FORGIVENESS

By Dave Denniston, CFA

D'oh! D'oh! D'oh!

Many residents slap themselves in the head continually trying to understand debt forgiveness programs and what they should (or should not) be doing.

It can feel like a hamster wheel turning round and round and round with no end.

In this chapter, we delve into debt forgiveness programs and lay out strategies for you to consider as you journey along your path.

SIX ASPECTS EVERY RESIDENT SHOULD CONSIDER WITH DEBT FORGIVENESS PROGRAMS

Due to the extremely high level of student debt that most physicians hold, many are eligible for several types of forbearance programs and debt-reduction programs. The difficulty lies in choosing among them all.

Truly, physicians have a wonderful opportunity to enroll in debt forgiveness programs. Later on, I'll ask you to think about and explore whether a loan forgiveness program may make sense for you. Here are a few factors that you may want to consider when looking over the possibilities:

> - **Does this cover my field of practice?**
> - **Do you specify a particular loan or can you get forgiveness on multiple loans?**
> - **Is this an employer or a state funded program?**
> - **Are the benefits taxable or not?**
> - **What is the length of the commitment?**
> - **Does the employer or the state pay down the loan each year or do they wait until the end of the commitment?**

Let's look at a couple of examples of some debt forgiveness programs...

THE PUBLIC SERVICE LOAN FORGIVENESS PROGRAM (PSLF)

The most common debt program that physicians look into is the 10-Year Public Loan Forgiveness program.

This is sponsored by the Federal Government and can cover virtually any field of practice. You don't have to specify a particular loan because it can cover all of your loans (assuming they have been Stafford, Perkins, and other federally backed programs). The benefits are currently not taxable, but this could change in the future.

As the name mentions, it is a 10 year program. The federal government will not forgive the balance until the end of the program.

HOW THE 10-YEAR PROGRAM WORKS

Here's how it works...

While you are employed full-time for a public service organization, you must make 120 on-time, full monthly payments (INCLUDES residency/fellowship).

Think about this for a minute- this is just seven years out of residency or maybe only three, four, or five years out of fellowship!

Note that if you have FFEEL and/or Perkins loans, you need to consolidate them into a Direct Consolidation Loan to take advantage of the program.

Qualifying employment is any employment with a federal, state, or local government agency OR a non-profit that has a 501(c)(3) status. Also, this includes certain non-profits that aren't 501(c)(3)s.

LET ME EMPHASIZE THIS STRONGLY- if you are employed by a hospital that has a non-profit 501(c)(3) status- you are probably eligible for this program!

Make sure to be aware whether the arm that you are working for is a non-profit or for-profit. Some non-profit hospitals can have a for-profit subsidiary for tax reasons.

Note that your monthly payments are substantially lower while in residency and fellowship. We will go through some examples later for after residency and fellowship.

Think about this for a minute....

If you are in residency for three years, you will only have seven years remaining on payments.

Meanwhile if you have a fellowship for three years in addition to three years of residency, you only have four years remaining on payments!

The bottom line is to make sure you to enroll AS SOON AS POSSIBLE while you are in residency and fellowship!

Here's how it works...

If you have FFEEL and/or Perkins loans, you need to consolidate them into a direct consolidation loan to take advantage of the program. This is a process will take one to three months to complete depending upon your situation.

OTHER DEBT FORGIVENESS PROGRAMS

STATE SPONSORED DEBT FORGIVENESS PROGRAMS

Besides, PSLF, there are some really exciting opportunities offered in every state.

Make sure to check out state sponsored programs at: https://services. aamc.org/fed_loan_pub/.

As of the time of this writing (July 2016), there were over 71 different programs available across the country!

Here is an example of a current program in Minnesota...

MINNESOTA URBAN PHYSICIAN LOAN FORGIVENESS PROGRAM

Who? Applicants are primary care medical residents, which include Family Practice, Obstetrics and Gynecology, Pediatrics, Internal Medicine and Psychiatry. You would apply July 1 to December 1 while completing medical residency training.

Requirements. Following completion of residency, the participant must plan to practice for at least 30 hours per week, for at least 45 weeks per year, for a minimum of three years in an underserved urban community.

The Nitty Gritty Payment Details. The state will re-pay up to $25,000 per year of service, not to exceed $100,000 or the balance of the designated loan, whichever is less.

Tax Consequences. These payments are exempt from state and federal income taxes. $25,000 is the taxable equivalent of $35,700 (assuming a 30% tax bracket).

Time Commitment. You must serve at least three years or otherwise must repay plus interest what they paid towards your loan.

Here's another example of another state forgiveness program…

OREGON PARTNERSHIP STATE LOAN REPAYMENT (SLRP)

Who? Applicants are primary care medical physicians and psychologists (among other non-physician positions).

Requirements. Must be a US Citizen. The application period is normally open during October and November each year and awards are made in December.

The Nitty Gritty Payment Details. Qualifying providers can receive a maximum award of $35,000 per year of 25% of total debt, whichever is smaller.

Tax Consequences. These payments are exempt from state and federal income taxes. $35,000 is the taxable equivalent of $50,000 (assuming a 30% tax bracket).

Time Commitment. Commit to service obligation of at least two years. One year extension may be awarded for up to three additional years, for a maximum service obligation of five years.

NATIVE AMERICAN FORGIVENESS PROGRAMS

Besides working for a private non-profit practice or a larger public entity or HMO, some physicians may want to consider another alternative- working with Native American Tribes.

You can learn more by going to: www.ihs.gov/loanrepayment/

Who? Applicants are physicians specializing in obstetrics/gynecology, psychiatry, internal medicine, family medicine, and pediatrics.

Requirements. Must serve at a location on a reservation or other specified place by IHS.

A few quirks to be aware of: IHS utilizes a ranking system to address the goal of filling staff vacancies in Indian health programs when granting LRP awards. This system assigns priority consideration to Indian health program sites with the greatest staffing needs in specific health profession disciplines.

Also, IHS gives priority to applications of American Indians and Alaska Natives and to individuals recruited through the efforts of Indian Tribes and Tribal or Indian organizations.

The Nitty Gritty Payment Details. Physicians are eligible to receive up to $20,000 per year in health professions educational loan repayment when working for the IHS.

Tax Consequences. These payments are subject to state and federal income taxes. IHS will pay an additional 20% to the IRS to offset increased tax liability.

Time Commitment. A two-year service commitment is required.

NATIONAL HEALTH SERVICE CORPS LOAN REPAYMENT PROGRAM

In addition to the state programs, there are various other granting national programs and opportunities. For example, the NHSC Loan Repayment program provides loan repayment assistance to licensed medical providers who serve in communities with limited access to health care.

There are both full-time and half-time options for service commitment. The dollar amount of assistance and length of service depend on participation in either the full- or half-time and on the need on the Health Professional Shortage Area (HPSA) score of the site.

Essentially, they are looking to fill physicians in "underserved" areas across the country. If you have one right in your area, this could be your ticket!

You can learn more about this program by going here: **http://nhsc. hrsa.gov/loanrepayment/**

Who? Selection is based on the staffing needs of the NHSC. For physicians, priority for selection will be given to those who have completed residencies in the following: family medicine, obstetrics/gynecology, pediatrics, psychiatry, geriatrics, or internal medicine.

Requirements. In exchange for loan repayment, participants are obligated to serve full-time upon completion of training at a designated NHSC-LRP site of their choice. US Citizenship required.

The Nitty Gritty Payment Details. Physicians may receive repayment of up to $50,000 in health professions educational loans (depending on site). Primary care providers working full-time at an NHSC-approved site with a HPSA score of 14 or above can receive up to $50,000 in loan repayment for committing to serve at site for at least two years.

Primary care providers working full-time at an NHSC-approved site with a HPSA score of 13 or below can receive up to $30,000 in loan repayment for committing to serve at the site for at least two years.

Tax Consequences. The loan repayments are exempt from gross income and employment taxes. These funds are not included as wages when determining benefits under the Social Security Act.

Time Commitment. It is a minimum of two years, but the physician could choose to stay longer. At the end of two years, Corps members can apply to continue their service and receive additional loan repayment. With continued service, providers may be able to pay off all their student loans!

FINAL THOUGHTS

There are so many kinds of debt forgiveness programs out there!

I know that if you are serious about being debt free that there is one out there for you.

In Napoleon Hill's slightly less famous book, *Think Your Way To Wealth,* he details the multiple interviews that he had with steel magnate and multi-billionaire Andrew Carnegie.

Mr. Carnegie stated that having a burning desire was a key ingredient to financial success.

My friends, the desire to be debt free MUST consume you.

Let it drive you to financial success & find several ways to eliminate your debts as soon as possible.

You can do it!

SOAK IT UP LIKE A SPONGE

There's a few physicians that I've spoken with that take the attitude, "Screw debt forgiveness programs, I want to take care of it myself!"

However, for many of us the prospect of paying back $300,000 or more in debt has us quivering in our boots.

If you refinance the debts, you are likely looking at payments upwards of $3,000/month and that's one hefty price to pay.

One of my podcast's past guests, Dr. Bo Liu, struggled with this decision. However, he decided to go ahead pursue PSLF.

He's made a few mistakes along the way (including selling some stocks at relative low value).

Find out why he pursued PSLF.

Learn more about Bo's journey at:
www.doctorfreedompodcast.com/bo

RESOURCES MENTIONED

State Sponsored Debt Forgiveness Programs:
https://services.aamc.org/fed_loan_pub/.

Native American Forgiveness Programs:
www.ihs.gov/loanrepayment/

National Health Service Corp Repayment Programs:
http://nhsc.hrsa.gov/loanrepayment/

Napoleon Hill's *Think Your Way To Wealth*
http://amzn.to/2ajp9ik- physical book
http://amzn.to/2amR98i - audio book

Physician Fireside Chat with Dr. Bo Liu:
www.doctorfreedompodcast.com/bo

CHAPTER 8

HOW IBR & PAYE CAN HELP YOU WIN THE GAME OF THRONES
By Dave Denniston, CFA

O kay, so we've laid out a few debt forgiveness programs.

Most of us reading this book probably won't use most of what we laid out in the previous chapter. I am guessing that you will go for PSLF (at least initially in your residency)!

If that is the case for you, let's dive into the first choice you'll need to make... which repayment program for PSLF should you consider and why?

HOW REPAYMENT WORKS

As you complete the Direct Consolidation Loan, you must pick a repayment program. The four most common programs are the Income-Based Repayment (IBR) Plan; the Pay-As-You-Earn (PAYE--or PER, as I sometimes refer to it) Plan; the Income-Contingent Repayment (ICR) Plan; and the 10-Year Standard Repayment Plan.

In this book, we focus on IBR and PAYE, as they require lower payments in residency and fellowship than do the other plans, which can lead to greater debt forgiveness.

After consolidating your debt, you start to make on-time monthly payments during the ensuing 120 months.

Make sure EVERY YEAR to complete, with your employer's certification, the Employment Certification form. You will also need to complete this form whenever you change jobs.

Submit the completed form to FedLoan Servicing (PHEAA), the PSLF servicer, following the instructions on the form.

FedLoan Servicing (PHEAA) will review your Employment Certification form, ensure that it is complete, and, based on the information provided by your employer, will determine whether your employment is "qualifying employment" for purposes of the PSLF Program.

DIFFERENCE BETWEEN INCOME-BASED REPAYMENT & PAY-AS-YOU-EARN REPAYMENT PLANS

The most common program is the Income-Based Repayment (IBR) plan. The second-most-common program is the Pay-As-You-Earn Repayment (PAYE) plan.

IBR and PAYE both accomplish the same goal: minimizing your student debt payments while you're in residency/fellowship, and then allowing you to pay back your student loans at a higher rate once you are making more dough.

REQUIREMENTS

Note that IBR and PAYE both require a "partial financial hardship."

This is different than a 10-year Standard Repayment Plan that you would automatically get forced into. Don't worry- you should be fine!

Not to confuse you, but if you are a resident and have a spouse with a similar income ($50k to $60k), you won't have any problems be considered to have a "partial financial hardship."

COMMITMENT

The commitment for IBR will be a monthly payment of 15% of discretionary income, whereas under PAYE the commitment will be only 10% of discretionary income. Note that discretionary income is defined quite specifically: it is your income minus whatever income meets poverty guidelines as established by the U.S. government.

ADJUSTMENTS

The federal government will ask questions about your household--i.e., spouse, spousal school loans, kids, etc.,--as these factors affect the poverty guidelines. How do they determine your income? By looking at your tax return!

This is an important distinction because the government is purely looking at your "adjusted gross income."

This means that they are taking a snapshot of your income AFTER pre-tax deductions for 401k/403b contributions, AFTER pre-tax deductions for health savings accounts, and AFTER deductions for any active business losses.

Also, this means that if you ended your residency/fellowship in June and started your first contract in July, then you would likely not have to start making higher payments UNTIL the following year.

For example, if you finished your residency in June 2017, then your higher payments would not take effect until past March or April of 2018 (or whenever you receive your renewal. It could be much later in the year).

However, the payments DO NOT take into consideration your overall student load debt, nor your age, nor whether you have a car loan, mortgage, etc. The student debt load is a particularly interesting factor to consider as we explore debt forgiveness programs.

Below is a table that I composed by entering information on the calculators at studentaid.ed.gov.

Note that I assumed that the person for whom I entered hypothetical financial information is married and has no kids and no spousal school loans; I assumed that the person's original loans were $20,000 to

$30,000 below the current loan amount; and I assumed that the loans carry an interest rate of 6.8%.

While you could likely easily qualify for IBR while in residency, the calculator on the website doesn't allow me to calculate the payment at a $200,000 income level, $150,000 loan amount for IBR.

However, we could safely assume that the payment should be $2,216/month, given the example below, because the monthly payment fluctuates with changes in compensation, but not with the loan amount.

Compensation	Loan Amt	IBR	PAYE
$150,000	$150,000	$1,591/mo	$1,061/mo
$150,000	$250,000	$1,591/mo	$1,061/mo
$200,000	$150,000	*Does not qualify	$1,478/mo
$200,000	$250,000	$2,216/mo	$1,478/mo

Note the tremendous difference between IBR and PAYE: over $500/month at the $150,000 compensation level and over $700/month at the $200,000 compensation level.

See how the IBR or PAYE amount does NOT change as the loan amount goes up? This is because the monthly repayment amount is primarily dependent on income.

There is one significant caveat as you choose between the two programs. To qualify for PAYE, you may not have any current student debt that originated before 2007.

How does all of this tie in with loan forgiveness programs? Let's take a look at an example of two physicians, Dr. Smith and Dr. Jones, each of whom started PSLF at the very beginning of residency. Each physician had an equal amount of student debt as he came out of medical school.

The Curious Case of Dr. Smith. Dr. Smith has been in residency for three years, has made 36 payments towards PSLF, and went right into practice. He is making $150,000 per year.

The Example of Dr. Jones. Meanwhile, Dr. Jones has also been in residency for three years, has made 36 payments towards PSLF, and has also just entered into practice. He is now making $200,000 per year.

Let's examine what would happen differently if each of them had enrolled in IBR or PAYE at the start of residency.

The table on the next page adds up the monthly payments from the previous example and multiplies the payments over seven years. There is no increase in salary. I'm keeping it simple and flat. The lifetime payments in the table are the combination of interest AND principal over those seven years.

Compensation	Loan Amt	IBR- Lifetime	PAYE- Lifetime
$150,000	$150,000	$133,644	$89,124
$150,000	$250,000	$133,644	$89,124
$200,000	$150,000	*Does not qualify	$124,152
$200,000	$250,000	$186,144	$124,152

At the end of the seven years (assuming continued employment at a qualifying non-profit), the remaining portion of each physician's debt would be forgiven.

For example, given $250,000 of loans and $150,000 worth of compensation, after seven years in practice you will have paid about $90,000 in PAYE relative to about $130,000 in IBR, assuming your taxable income is $150,000.

After Dr. Smith completes 84 remaining payments, now that he is in practice, this would be approximately $225,000 worth of forgiveness with IBR, versus $265,000 of forgiveness with PAYE.

This is why PAYE is superior to IBR when student debt forgiveness programs are tied in with PAYE.

Additionally, the higher your loans, the more beneficial it will be to enroll in PAYE.

Let's say that you have $250,000 in student loans.

Consider this: at an interest rate of 6.8%, you are accruing interest of about $17,000 annually, or $1,416/month. With PAYE, you would have been paying $1,478/month, barely paying off any principal.

Then, over 7 years, if you are enrolled in the PAYE program, you will have paid about $124,000 and will have debt forgiveness of almost $250,000 of principal--and all of these dollars will likely be tax-free!

Giving up this financial gift is the tax equivalent of almost $360,000, or $51,000/year over 7 years, assuming a 30% tax bracket.

Even with IBR, you would still have debt forgiveness of nearly $200,000 or the approximate pre-tax equivalent dollar value of $285,000.

Either program is wonderful, but PAYE is better for debt forgiveness purposes!

Remember: as we mentioned earlier, in order to qualify for PAYE, you must have student debt that originated after October 2007. This fact will likely begin to affect residents and fellows who started residency in 2015, and will affect even more graduating residents and fellows over the next few years.

ATTENTION: YOUNG PHYSICIANS WHO WANT TO GROW THEIR FAMILIES

Tick… tick… tick…. The biological clock may be ticking.

I was meeting with a wonderful couple the other day, two neurologists. They are just transitioning into practice and they are so excited!

Their income is going to be awesome and they'll have so much cash flow that they won't know what to do with it.

Okay, okay, I'm sure they'll figure out some ways to spend that dough. =-)

In reviewing their financial situation, we came to the tough question of debt.

He has $300k+ in medical school debt and she has $200k+ in medical school debt.

Here they are: they have gone through undergrad, gone through medical school, gone through residency, and have now gone through fellowship. They are in their mid-30s and have been left with a mountain of debt.

<u>Time is running out! Tick… tick… tick… Her biological clock is ticking and they want to start a family.</u>

They both have enrolled in IBR and have been minimizing their payments, yet they are checking off the days towards the time when PSLF will allow them to be debt-free.

However, when we talked about their future plans and their family, I wondered whether or not debt forgiveness through PSLF is the right approach.

You see, after having their first kiddo, she is mulling over cutting back down on work to half-time, a 0.5.

I thought that was great and wonderful, but it doesn't work for IBR & PSLF!

PSLF requires that you work FULL-TIME in order to claim your credit for the year. On studentaid.ed.gov they say:

"You are generally considered to work full-time if you meet your employer's definition of full-time or work at least 30 hours per week, whichever is greater.

If you are employed in more than one qualifying part-time job at the same time, you may meet the full-time employment requirement if you work a combined average of at least 30 hours per week with your employers."

This is very interesting: you have to meet your employer's definition of full-time OR you must work at least 30 hours per week, whichever is **GREATER**.

In this case, half-time at 20 hours or 25 hours per week isn't going to cut it. Those years that she takes to continue working, BUT not full-time, in order to raise her family, will NOT qualify!

Thus, I was strongly pushing them towards counting the husband-neurologist's loans towards IBR/PSLF, as planned. After all, we are talking about $300k+ in potential forgiveness!

However, we need to change it up for the wife-neurologist's loans. I emphasized that we should refinance her loans through SoFi, DRB, or another loan consolidator.

I ended up listing off five different reasons why I thought this was important:

1. She was considering going half-time.
2. This ensures that no matter what happens with the government and PSLF, about half of their family's loans are under their own control and thus can be erased on their terms.
3. It immediately lowers the interest rate on her loans.
4. Once they have their debts paid off, it allows her to have the freedom to quit altogether, and to stay home with the kids if that is what they decide is best.
5. This makes an unknown variable a known variable, and that will make it easier for them to plan.

You should have seen the look they gave me.

Their noses crinkled, eyes went blank, and heads dipped downward. The room was very quiet.

They struggled with the idea, and to be honest it took two years, but they did implement this strategy.

I worry about doctors like these two neurologists, and can audibly hear the clock going tick... tick... tick...

WHAT CAN WE CONCLUDE FROM IBR VS. PAYE?

If you are working for a **non-profit** entity, PAYE is probably the better option unless you do not qualify due to the origination of your student debt. If you are ineligible for PAYE, IBR will still be a fine choice.

I would strongly suggest that you NOT put extra payments towards your debts if you are enrolled in the 10-Year Public Loan Forgiveness

Program, unless you think that you may not be ready to make a 10-year commitment to staying in the non-profit community.

If you are currently working for a **non-profit** and are considering **transitioning to a for-profit** practice after residency (and maybe you're not sure where you are going to be headed), <u>IBR</u> would be my recommendation.

Keep in mind that you can make extra payments beyond the minimum that IBR requires, and this will pay off your obligation sooner, once you are in practice.

TWO DOCTORS: MARRIED... BUT SEPARATED

<u>What if you are married? Is there a different strategy?</u>

Let me take you back a few years...

It's the happiest day of your life! Your wedding day... Best. Day. Ever.

Smiles, hugs, the white gown, the champagne, the first dance, the cake, and the limo. Laughter peals through the air. Fleeting memories of vows and eager faces, music hanging in the air.

Although, you could do without the bit of the mother-in-law drama!

You're a physician who is married to another physician. There is no doubt: the two of you have an incredibly bright future ahead of you. After all, you're going to be making almost $400,000 combined!

You're just getting through the final year of your residency. Practice is right ahead! You'll finally be able to buy a house, go on vacations, and not be scraping by on a champion's diet of apples, oranges, and Top Ramen.

Then it hits you like you just ran into a brick wall: $500,000 of medical school debt... How the heck are you ever going to get out of that debt?

Is it going to take 20 years? 30 years?

You've enrolled in the Public Service Loan Forgiveness Program (PSLF) and you think it may take only another 8 years, if you are lucky. However, the payments are going to be killer in another year--almost $5,000 a month between the two of you! That will de-rail your hopes of saving for a home quickly. How the heck could you afford it?

Then, you meet some crazy financial guy who tells you that you need to get separated...

SAY WHAT?

We just got married and now we have to get separated?

WHY TWO MARRIED PHYSICIANS SHOULD FILE AS "MARRIED FILING SEPARATELY"

Okay, I don't mean that these two physicians have to be separated legally; instead, I mean that they should remain married, but should be filing THEIR TAXES separately.

The two statuses ("real-life" marriage and "tax-status" marriage) currently do not have to be one and the same. It is a choice!

This choice came up recently when two physician clients came into my office to explore their options on re-paying their debts.

DISCLAIMER: These are two physicians who are working in a hospital setting, thus under a non-profit. Remember, in order to qualify for loan forgiveness under PSLF, you have to work for a non-profit or for a government entity.

We explored the differences between *married filing jointly* versus *married filing separately*. I was astounded by the results.

Below are three tables showing several scenarios that we ran. These three tables show how *married filing separately* versus *married filing jointly* can affect your PSLF payments, and thus the potential loan forgiveness.

In this real-life scenario, the wife is now in practice. She transitioned in July of 2014. He is two years behind her. He transitioned to practice in July of 2016.

At the time we discussed these scenarios, it was late 2014.

Keep in mind that the IBR payment changes AFTER you report your income--let's say by May of each year. For purposes of simplicity, we will assume that the change happens in January of each calendar year.

the wife---Dr. Giselle Smith--$110,000 in eligible student debt

Scenario	Income	IBR Payment
Single	$55,000	$469/mo
Married Filing Jointly/ One Still in Residency, Other in Practice	$240,000	$726/mo
Married Filing Jointly/ Both in Practice	$370,000	$1,100/mo
Married Filing Separately (Only Her Income)/ Transition Year	$120,000	$328/mo
Married Filing Separately (Only Her Income)/Fully in Practice	$185,000	$537/mo

the husband--Dr. Tom Smith--$315,000 in eligible student debt

Scenario	Income	IBR Payment
Single	$53,000	$444/mo
Married Filing Jointly/One Still in Residency, Other in Practice	$240,000	$2,002/mo
Married Filing Jointly/Both in Practice	$370,000	$3,220/mo
Married Filing Separately (Only His Income in Residency)	$55,000	$349/mo
Married Filing Separately (Only His Income)/ Transition Year	$120,000	$953/mo
Married Filing Separately (Only Her Income)/Fully in Practice	$185,000	$1,557/mo

Combined--Husband and Wife Together

Scenario	IBR Payment: Married Filing Jointly	IBR Payment: Married Filing Separately
Year 1: She is fully in Practice (transition year income was previous year), He is full year in residency	$2,084/mo	$677/mo
Year 2: She is fully in practice, he is full year in residency	$2,728/mo	$886/mo
Year 3: She is fully in practice, he is in transition year	$3,074/mo	$1,490/mo
Year 4 & beyond: Both are fully in practice	$4,320/mo	$2,094/mo
TOTAL PAYMENTS IN 4 YEARS	$146,472	$61,764
DIFFERENCE	$84,080	

We could go on and on, but check it out: You could be saving yourself literally HUNDREDS OF THOUSANDS of dollars IF your debts are forgiven through PSLF, AND if you take advantage of the option of being married but filing your taxes separately.

Look at this decision holistically before jumping into the pool and using this tax-filing method. There are real-world tax consequences to this decision when you file your taxes as married filing separately versus married filing jointly.

In this particular instance, the difference was minimal--only about $2,000 per year in combined federal and state income taxes. The benefit of the *filing separately* tax status on their debt reduction far outweighed the cash-flow hit that the couple took on their federal and state taxes.

Make sure to discuss any potential tax impact with your advisors so that you fully understand the consequences.

THE DEBT-LADEN PHYSICIAN

Every so often, I offer a free 30-minute strategy session to physicians, as a way to thank them for all they've done for our family and for people who are part of my community. I was heartbroken when I spoke with a physician from Oregon recently. You see, he had nearly $400,000 in student debt.

Unfortunately, he is in practice and no PSLF is available, because he is an ER physician and is working at a for-profit. He is paying $1,500/month, but that isn't even covering the interest on his debt!

We strategized a few solutions together--none of them are easy—solutions such as moving to a different state (like Alaska, a state that pays twice as much), such as picking up additional shifts doing locums, or such as shifting to a job at a non-profit.

I even directed him to save LESS, so that he could pay down debt MORE. By the end of our time together, he had direction, where previously he had doubt.

Interested in being a case study & being coached on the podcast? You can apply by e-mailing me at dave@doctor-freedompodcast.com

RESOURCES MENTIONED

The Employment Certification Form:
https://studentaid.ed.gov/sa/sites/default/files/public-service-employment-certification-form.pdf

Fed Loan Servicing:
http://www.myfedloan.org/

Instructions for Employer Certification:
https://studentaid.ed.gov/sa/sites/default/files/public-service-employment-certification-form-instructions.pdf

Apply To Be A Podcast Guest/ Case Study
dave@doctorfreedompodcast.com

CHAPTER 9

THE CRAZY WORLD OF RE-PAYE
By Amanda Liu, MD

(Note: The majority of this chapter first appeared as a guest post on The White Coat Investor and is used with permission at: http://whitecoatinvestor.com/tag/is-repaye-right-for-me/)

As you have read through these last few chapters, you may be scratching your head and may have been confused about all these options!

There's PSLF, IBR, PAYE, and refinancing my loans. Which of these should I choose?? Which is the right fit for me??

You may even have lost a few of your hairs as you are stressing and unsure of where you should turn.

Well, my friends… there is one more option that you should be considering.

However, before you go on further, let's take a break for a minute…

…

….

At the very end of this chapter, I (Amanda) am going to give you access to an assessment. This assessment will help you wade through the waters and determine what repayment options may be right for you!

Make sure to turn to page 87 to discover how to find this assessment.

Anyhow, as we mentioned, there is another option…

It's Revised Pay as You Earn (REPAYE). REPAYE is the new kid on the block of government income-driven repayment plans. In this chapter, I (Amanda) summarize the principles and then review eight different case studies illustrating them.

These three principles illustrate the main differences between REPAYE and its friends, IBR and PAYE.

1. If your monthly payment doesn't cover the interest that accrued, the government pays:

    1. In REPAYE, 100% of the difference on subsidized loans for the first 3 years, then 50% of the difference thereafter.
    My friend and fellow physician blogger, Dr. Bo Liu, at FutureProofMD.com noted that subsidized loans are usually around $50,000 of the total debt.

    2. In REPAYE, 50% of the difference on unsubsidized loans indefinitely.

3. In comparison, NO interest is subsidized in IBR or PAYE

2. You pay monthly installments equal to 10% of your household discretionary income (counts your spousal income, no matter how you file taxes; you CANNOT file as "married filing separately," thereby reducing your monthly payment as in IBR and PAYE)

3. No payment cap (especially important after residency completion, as REPAYE can erase potential benefits of PSLF)

QUESTIONS

When deciding about REPAYE, you need to know the answers to the following questions:

1. What is my TRUE REPAYE interest rate after the subsidy?
2. What are my interest rate/term offers for refinancing?
3. Do I plan on pursuing PSLF?
4. Do I plan on pursuing (taxable) forgiveness through IBR, PAYE, or REPAYE after 20-25 years of payments?
5. Am I married to a spouse making an income similar to or greater income than mine?

CASE STUDIES

First we'll look at six "flow-chart" case studies, and then we'll consider two reader case scenarios, with step-by-step evaluations of REPAYE. Assumptions for the flow-chart case studies are as follows:

1. Class of 2016
2. $200K total federal student loans
3. 6.1% weighted interest

4. Monthly accrued interest: $1017

5. REPAYE monthly payment requirement calculated with repayment estimator at: https://studentloans.gov/myDirect-Loan/mobile/repayment/repaymentEstimator.action#view-repayment-plans

6. Flow-chart case studies include the following:

   o Single with med school & moonlighting

   o Single without med school or moonlighting

   o Married, household of 5, with some spousal income

   o Married, household of 2, with some spousal income

   o Married, household of 5, with high spousal income

   o Married residents, household of 2, with some moonlighting income

CASE STUDIES 1 AND 2: THE SINGLE DOCS

Case Study 1: a single PGY1 with med school income ($25k) and moonlighting income ($35k). His PGY1 REPAYE payments are based on his 2015 tax return, which showed $25k AGI (Adjusted Gross Income).

His PGY3 REPAYE is based on his 2017 tax return, which shows $100k income. With increasing AGI from tax returns 2015 to 2017, his REPAYE payment and effective REPAYE interest rate both increased.

In Case Study 2, a single PGY1 had no income in medical school but did do some moonlighting in residency. His PGY1 REPAYE payments are based on his 2015 tax return, which had $0 AGI; his PGY3 REPAYE is based on his 2017 tax return, which shows $100k income. His REPAYE payments and effective interest rate were lower than were those in the Case Study 1, but also increased over time as his income increased. His initial effective interest rate is exactly half of his weighted interest rate.

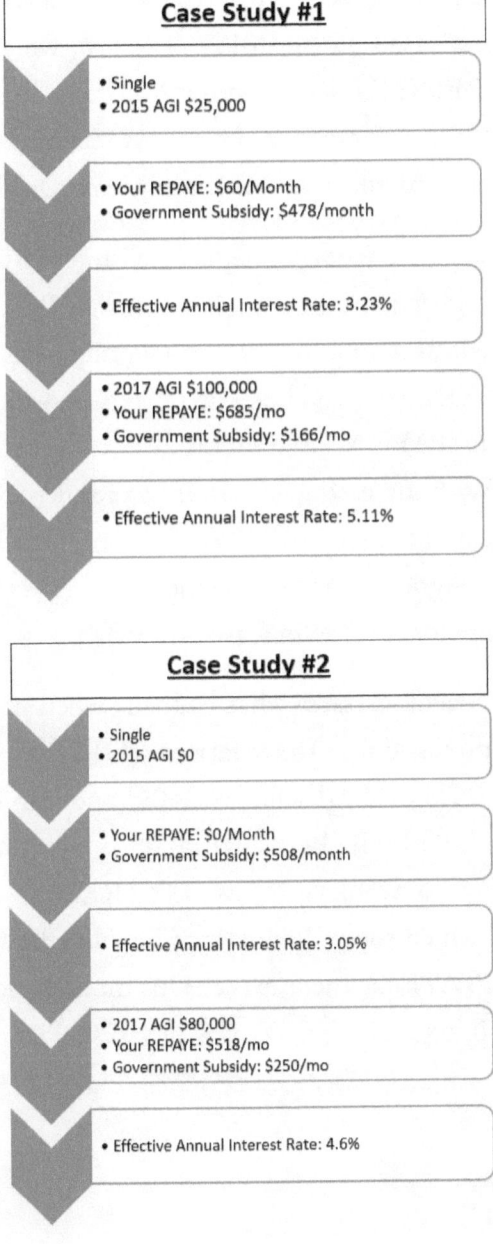

Case Study #1

- Single
- 2015 AGI $25,000

- Your REPAYE: $60/Month
- Government Subsidy: $478/month

- Effective Annual Interest Rate: 3.23%

- 2017 AGI $100,000
- Your REPAYE: $685/mo
- Government Subsidy: $166/mo

- Effective Annual Interest Rate: 5.11%

Case Study #2

- Single
- 2015 AGI $0

- Your REPAYE: $0/Month
- Government Subsidy: $508/month

- Effective Annual Interest Rate: 3.05%

- 2017 AGI $80,000
- Your REPAYE: $518/mo
- Government Subsidy: $250/mo

- Effective Annual Interest Rate: 4.6%

CASE STUDIES 3, 4, AND 5: MARRIED WITH CHILDREN

In Case Study 3, we have a married PGY1 with spousal income since medical school ($30k) and moonlighting income ($15k). His PGY1 REPAYE is based on his 2015 tax return, which showed $30k AGI; his PGY3 REPAYE is based on his 2017 tax return, which shows $110k income. With increasing AGI from tax returns 2015 to 2017, his REPAYE payment and effective REPAYE interest rate both increased.

In Case Study 4, we see what happens with fewer family members and somewhat higher income. This is a married PGY1 with spousal income since medical school ($50k) and moonlighting income ($15k). His PGY1 REPAYE is based on his 2015 tax return, which showed $50k AGI; his PGY3 REPAYE is based on his 2017 tax return, which shows $130k income. With increasing AGI from tax returns 2015 to 2017, his REPAYE payment and effective REPAYE interest rate both increased. Note how much higher his initial payment and effective interest rate are than they were for the doc in Case Study 3.

Case Study 5 demonstrates what happens when you're married to a high-income professional. This married PGY1 has enjoyed a high spousal income since medical school ($150k) and also has had moonlighting income ($15k). His PGY1 REPAYE is based on his 2015 tax return, which showed $150k AGI; his PGY3 REPAYE is based on his 2017 tax return, which shows $230k income. Note how, by the end of residency, his REPAYE payment exceeds his monthly accrued interest, and his effective interest rate is 6.1%. (He receives $0 of interest subsidy.)

Case Study #3

- Married: Household of 5
- 2015 AGI $30,000

- Your REPAYE: $0/Month
- Government Subsidy: $508/month

- Effective Annual Interest Rate: 3.05%

- 2017 AGI $110,000
- Your REPAYE: $561/mo
- Government Subsidy: $228/mo

- Effective Annual Interest Rate: 4.73%

Case Study #4

- Married: Household of 2
- 2015 AGI $50,000

- Your REPAYE: $216/Month
- Government Subsidy: $400/month

- Effective Annual Interest Rate: 3.70%

- 2017 AGI $130,000
- Your REPAYE: $883/mo
- Government Subsidy: $67/mo

- Effective Annual Interest Rate: 5.7%

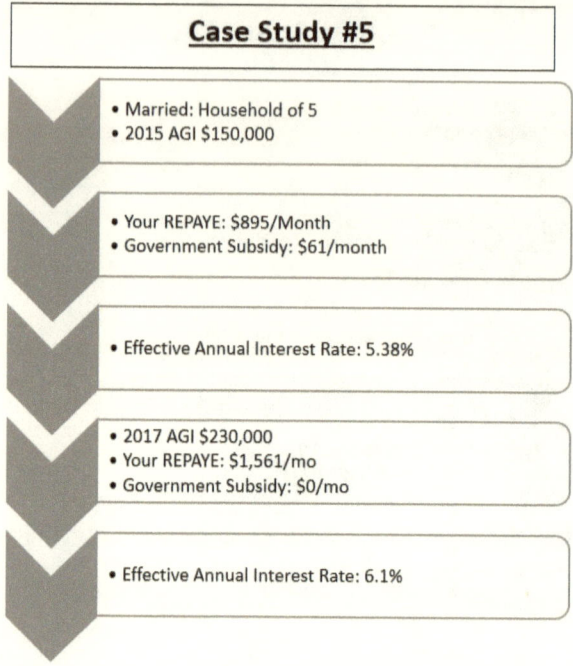

Case Study #5

- Married: Household of 5
- 2015 AGI $150,000

- Your REPAYE: $895/Month
- Government Subsidy: $61/month

- Effective Annual Interest Rate: 5.38%

- 2017 AGI $230,000
- Your REPAYE: $1,561/mo
- Government Subsidy: $0/mo

- Effective Annual Interest Rate: 6.1%

Let's take another break for a moment and focus on what we've observed so far...

- The lower your income, the more beneficial REPAYE is to you.
- The sooner you start PSLF in residency, the more beneficial REPAYE is to you.
- The larger your household, the more beneficial REPAYE is to you.
- The higher your debts, the more beneficial REPAYE is to you.

<u>Just imagine if a resident had $300,000 of student loans rather than $200,000.</u>

That's approximately another $6,000 a year of interest that the resident would be accruing on their loan balance if they were using IBR or PAYE.

With REPAYE, the government will subsidize higher and higher portions of your interest, according to the level of your debt.

CASE STUDY 6: MARRIED RESIDENTS

In Case Study 6, we have two docs, a married PGY1-PGY1 couple with $0 medical school income and a little moonlighting income in residency ($15k each). Their PGY1 REPAYE payment is based on their 2015 tax return, which showed $0 AGI; their PGY3 REPAYE is based on his 2017 tax return, which shows $130k in income.

Check out the jump in effective interest rate as they start moonlighting.

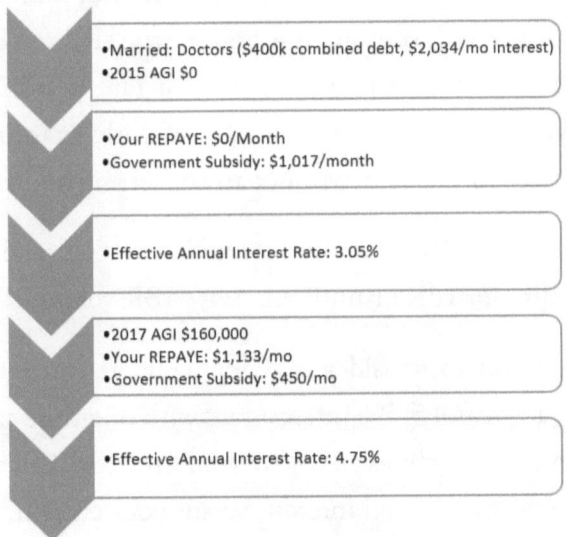

- Married: Doctors ($400k combined debt, $2,034/mo interest)
- 2015 AGI $0

- Your REPAYE: $0/Month
- Government Subsidy: $1,017/month

- Effective Annual Interest Rate: 3.05%

- 2017 AGI $160,000
- Your REPAYE: $1,133/mo
- Government Subsidy: $450/mo

- Effective Annual Interest Rate: 4.75%

CASE STUDY 7: MARRIED TO AN ATTENDING AND IN IBR

"I have $150k student loan principal, $50k interest, all at 7%. My spouse makes $200k as an attending. We are a family of 6. I make $60k as a PGY3/6. Should I switch to REPAYE?"

First, calculate your monthly REPAYE payment. Is it affordable? Using the online calculator, you can see that your REPAYE monthly payment is $1,759/month.

You currently have $200k total debt (all consolidated): $150k principal and $50k interest. Your current IBR monthly interest is calculated on the $150k principal. But when you leave IBR to go REPAYE, your $50k interest will capitalize, and your new principal will be $200k.

Your REPAYE monthly interest is calculated on the new 200k principal:

$200k*7%= $14k annual interest without subsidy. Divide that by 12 to get your monthly interest of $1,167. Since your monthly payment is greater than the monthly interest you accrue, REPAYE will NOT subsidize you. This means that your interest rate is still 7%. So you should not switch to REPAYE, since it ONLY works against you by capitalizing your current interest, without any offsetting interest reduction.

If you stick with IBR, your current interest won't capitalize until you no longer qualify for IBR or until you leave IBR voluntarily.

Another option to consider is refinancing. If you refinance, you will likely get a 7-year 4.5-5% interest rate with one-time capitalization when the loan company buys your loan from the feds. This will give you $200k*4.5% = $9k in annual interest, versus your current $150k*7% = $10.5k interest. This isn't a whole lot of savings, but you may also benefit from a lower $0-$100 payment. Obviously, refinancing ensures that you will not receive PSLF.

[White Coat Investor's Note: Some docs in this situation, especially if going for PSLF, may wish to file their taxes as Married Filing Separately and to use either the PAYE (if they qualify) or the IBR programs.]

CASE STUDY 8: MARRIED TO A SPOUSE WITH NO INCOME AND IN IBR

"I have $150k student loan principal, $50k interest at 7%. My spouse stays at home. We are a family of 5. I make $60k as a PGY2/6. Should I switch to REPAYE?"

Per the calculator, your REPAYE monthly payment is $145/month. This means you will enjoy some interest savings by going to REPAYE. To switch from REPAYE to IBR, there's one middle step. They will put you on standard 10-year repayment, where you need to make at least one payment before you can select REPAYE. If the standard 10-year payment is too high, you can apply to make a reduced payment. Another factor to consider is that when you leave IBR (with the goal of going to REPAYE), your interest will capitalize.

You currently have $200k total debt (all consolidated): $150k principal and $50k interest. Your current IBR monthly interest is calculated on the $150k principal. But when you leave IBR to go REPAYE, your $50k interest will capitalize, and your new principal will be $200k.

Your REPAYE monthly interest is calculated on the new $200k principal, so $200k*7%= $14k annually or $1,167 monthly not counting the REPAYE subsidy. Subtract your monthly payment from your monthly interest

= $1,167 -$145 = $1,022. REPAYE will pay 50% of $1,022 = $511. Your net annual interest after subsidy will be

12 x ($1,167 monthly interest -$511 monthly subsidy)

= $7,872.

Thus, your effective interest rate will be $7,872/$200,000 = 3.94%. Your total interest saved will be $6,132/year as annual simple interest.

(Technically, there will be a little more savings, since the feds calculate interest daily, not annually.)

You could also compare this to refinancing during residency. Let's say you can refinance into a 7-year loan with a 5% to 6% interest rate. This is the most common rate we see for residents. In comparison, practicing physicians are getting 3.5% to 5.5% depending upon income, credit score, etc.

Obviously 5% is more than 3.94%, so you'll end up paying more in interest if you refinance while in residency. That may change as you progress through residency. If you're not going for PSLF, you may wish to recalculate your effective interest rate each year and compare that rate to what you can obtain as a refinance rate.

Although, here's the kicker. If you choose to refinance your loans at a private lender later, all that subsidized interest gets added back.

The lesson here: don't look at REPAYE as a temporarily solution. Be committed to it. That capitalized interest could come back to haunt you!

Otherwise, you are likely better off in IBR, PAYE, or simply refinancing your loans if you think your plans are likely to change.

TWO CAVEATS

Remember that consolidating your loans will RESET your PSLF 120-payment clock. If you have already consolidated your loans or if all of your loans are PSLF eligible (that is, they don't require consolidation), then you don't need to worry about this.

Also, remember that while it is true that under current law you can switch from REPAYE back to IBR upon residency graduation (to avoid the potentially higher payments under REPAYE, since those payments aren't capped like IBR and PAYE payments are), that law could change.

SUMMARY

Compared to IBR or PAYE, REPAYE helps those with little household income and a large debt, because the interest subsidy only exists on the gap between your accrued interest and your payment.

The more you pay, the smaller the gap between your monthly payment and the monthly accrued interest, so the less interest subsidy you benefit from. When you pay exactly the amount of your monthly accrued interest, or else a greater amount, you get no subsidy at all.

For a single physician that has a long residency/fellowship, REPAYE could be an amazing fit.

For a married physician with a shorter residency, they are likely better off with IBR or PAYE.

Alternatively, for a physician that may wish to refinance their loans and pay them back using a private lender, REPAYE could be a good short-term stop gap between residency and transition to practice.

What do you think?

Have you switched from IBR or PAYE to REPAYE?

Are you confused about what to do?

Take our quiz at www.doctorfreedompodcast.com/debtquiz.

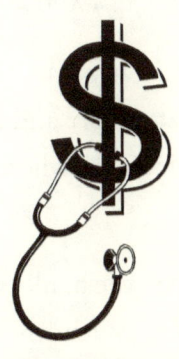

CHAPTER 10

WHAT IF... I DON'T WANT (OR CAN'T DO) A DEBT FORGIVENESS PROGRAM?

By Dave Denniston, CFA

You may be wondering... *What if I don't work for a non-profit?*

You may be wondering... *Why couldn't I just pay off my debt quicker on my own?*

When the banks were thrown into the pits in the depths of the debt crisis, we saw a tremendous change in the way student loans operated.

Practically every single resident or fellow is paying somewhere between 6% and 7% in interest. On $200,000 worth of debt, we're talking about $13,000 a year in interest--or nearly $1,000 a month!

If you think about virtually any loan, like a mortgage or a car loan, if you have a high interest rate today and if interest rates drop tomorrow, you can refinance your loan elsewhere. Your 6% loan can become a 5% loan.

Yet, for student loans, when the banks stepped away and the federal government became the primary lender- there became this vacuum. You couldn't refinance student loans. Docs have been stuck with these crazy high interest rates even though the Fed has driven interest rates to be crazy crazy low. And we couldn't make refinancing high interest rate student loans happen!

In this void, private equity and some small banks have started to step in and to make refinancing possible today for many physicians.

TWO EXAMPLES TO LEARN HOW TO SLASH YOUR INTEREST RATES!

SoFi (Social Finance) backed by private equity investors became one of the early entrants into this space.

In a podcast on DoctorFreedomPodcast.com, founder Dan Macklin said, "I was at Stanford Business School... It's one of the best business schools in the country. These are very smart people, very employable people, but we saw that our classmates were paying really, really high rates for their loans. First, a lot of people were borrowing and secondly, they were paying ridiculously high rates for their loans: six, seven, eight percent. We thought it was strange that once you graduated you couldn't then refinance that debt."

(You can find the whole interview with Dan Macklin at www. doctorfreedompodcast.com/sofi .)

Today, SoFi specializes in refinancing loans AFTER a physician has transitioned to practice. They currently do NOT have a program for which a physician in residency is eligible.

Paying. Most physicians are paying 6%, 7%, or even 8% on student debt. SoFi's rates have varied from as low as 3.5% on a fixed-rate loan to 1.9% on a variable rate.

Saving. Physicians could save 2%, 3%, 4%, or 5% and could slash their costs in half, saving tens of thousands of dollars in the process.

How It Differs. The interest rates quoted by SoFi can be different from physician to physician. Dan Macklin noted that *"[t]here's no exact debt-income ratio that we're looking for. There isn't a perfect number that I can tell you... It's slightly different for everybody because we look at a number of things including the credit history, what school you went to, what kind of profession you're doing, where you major, etc. But among those things the salary is very important."*

No Cost. The amazing fact is that regardless of whom you choose--whether it is SoFi or DRB or another company--there is NO cost to doing a refinancing. With mortgages, I usually think of mortgage refinancing and points.

Yet, there are NO origination fees, refinancing costs, or points currently associated with the loan refinancers. There's also no repayment penalty. So, you could pay back your loans early if you desired to do so.

How many people can qualify for SoFi's lowest interest rate?

Dan Macklin said, *"It's pretty much a bell-shaped curve in terms of our rates. So, if I take our five-year rate... it goes from 1.9% to 4.1%. The average person that gets approved by SoFi [is] somewhere in the middle.... Physicians [are] among the kind of a demographic who many of them get that [lowest] rate, not everybody gets there of course, but a very, very high number of people do get it."*

You can get a $300 refinancing bonus by going to SoFi.com/PoF to refinance your loans with SoFi.

Another company that specializes in refinancing medical school loans is Laurel Road (formerly DRB). They started a little bit later in the debt refinancing game than SoFi, but are aggressively moving into the territory.

Laurel Road has very similar interest rates and offers to those of SoFi. However, they did note in a recent interview on DoctorFreedomPodcast. com that they DO NOT base their rates on a bell-shaped curve, and instead tend to offer lower competitive rates to many physicians.

<u>In addition, Laurel Road recently launched a program that allows residents to refinance their loans WHILE they are in residency or fellowship. This option is unique to the marketplace.</u>

Laurel Road has tried to imitate the Public Service Loan Forgiveness Program (PSLF) and income-based repayment program (IBR) as closely as possible, by limiting the maximum monthly payment to $100/mo while physicians are in residency.

In the meantime, with interest rates being locked in at historic low rates, your interest is likely accruing at half of what the amount used to be. Rather than interest of $1,000 a month, you may owe only interest of $800 or $900 a month.

As noted in the previous chapter, the rates for residents are 1% to 2% higher than practicing physicians in this program.

This historical low in interest rates and the corresponding low accumulation of monthly interest owed is tremendous when you consider the value (or, if you are in debt, the COST) of compounding!

The bummer is that when you refinance your debts with a private company like SoFi or Laurel Road you can NO LONGER participate in PSLF. No debt forgiveness, only debt pay-off that you create for yourself!

However, you'll be in control of your own destiny rather than relying on the government, which is an advantage worth considering.

You can find a whole podcast interview with Alex Macielak at www.doctorfreedompodcast.com/drb.

WHAT TO DO IF YOU ARE A RESIDENT OR A FELLOW

Companies such as LinkCapital, Laurel Road, and a brand-new player, GradSchoolLoans.com, will refinance your debts while you're in residency.

Now, I will tell you that the rates offered by these companies are higher as a resident than when you are practicing as a physician.

As I'm writing this chapter, those companies are offering around a 5% to 5.5% interest rate.

So, I'm a little disappointed by that.

But, if you're looking to refinance debts, you know that you want to pay them off...

Why not do Laurel Road? Why not go with LinkCapital? Why not go with GradSchoolLoans.com?

This way, you're saving yourself at least 1% on that debt. Think about that, my friends...

On a $250,000 pile of student loans... That's the equivalent of $2,500 a year in interest, at least. Think about that: that's $200 a month you're saving yourself by lowering your interest rate TODAY, while you're in residency. You must!

What's nice about these companies is that your minimum payments are small while you're in residency.

I think Laurel Road's payment is around a $100 a month. LinkCapital's is around $100 a month.

GradSchoolLoans.com is doing it for about a dollar a month.

Now, the thing that I like about all these programs is this: Let's say we fast-forward a few years:

What if rates are still low when you come out of residency?

The awesome news is that there are **no prepayment penalties** for refinancing with another company in the student loan world (or even for refinancing within the existing company or asking them to take another look at your loan).

This is different than the case for a mortgage or a car loan. Some of those big bad banks & finance companies want your money and your loan soooooo badly.

Why? They REALLY, REALLY want to collect that interest, and *they will charge you a penalty if you refinance the loan somewhere else* (even if you pay off the loan early as a good borrower).

Luckily, this is NOT the case with the student loan refinancing companies that we are talking about.

Let's break it down a little bit:

Laurel Road is backed by an actual bank: Darian Rowayton Bank. So they typically hold onto the loans they originate.

In comparison, LinkCapital and GradSchoolLoans are funded by venture capital, private equity, and angel money. They are in the game of turning loans around and making a quick buck on their investment.

More than likely, these latter groups are going to sell your loans to someone else. I'm not really sure that this matters, because, luckily, you maintain the servicer: the people to whom you send the payment typically WON'T change.

Also, when you contract for a fixed loan, it is fixed. It's not like they are suddenly going to have the ability to jack up the rate on you.

What does this all mean?

At the end of the day, **it's all about who gives you the best rate.**

Who will give you 5%? Who will give you 5 ½%?

It doesn't matter! Go for the one who will give you the lowest rate.

One of the questions I often receive is this:

"Does it hurt you to apply to multiple companies? Will it hurt your credit score much?"

Consider this... one of the components of a credit score is: how often has your credit been pulled?

If it has been pulled multiple times--think five, six, seven, eight, or nine times—then, yes, too many applications could hurt your credit score. However, if you keep the number of refinancing applications to three, you'll be fine.

Most of these companies do a "soft" credit pull, which should have minimal effect on your credit score.

WHAT TO DO IF YOU ARE A PRACTICING PHYSICIAN

Now, let's say you're already in practice.

This list gets longer! There are quite a few companies that would happily take over your student loans from the federal government.

Laurel Road and LinkCapital are certainly still in the mix. But there's also SoFi, CommonBond, Earnest, and many more.

It has been my observation lately that SoFi and Earnest tend to be the best two individual companies that I see for physicians in practice.

Going Public. SoFi has been on the verge of going public. They have been backed by venture capital in the past. However, they are growing bigger and bigger and bigger and are becoming a bank. They are offering mortgages, car loans, and personal loans, as well as some other bank-like products. Don't be surprised to see them create savings accounts and money markets soon.

The Bait. These companies lure you in with great rates for student loans, and they know that they can offer you other products (like mortgages or personal loans). I personally believe that they are trying to attract as many physicians as possible, both for now and in order to look good to the public markets prior to their IPO.

They Make What They Keep. Also, in this same spirit, I'm pretty sure that SoFi is now KEEPING most of their physician student loans. They are no longer selling them as often as they did previously. This does make things pretty seamless.

In comparison, Earnest is much, much smaller. However, they are also backed by venture capitalists (VCs). They mirror LinkCapital & CommonBond pretty closely, but they don't deal with residents.

Anywho... I've seen a few cases now where both of them, SoFi and Earnest, are going to be neck-and-neck, at about 4%, 4 ¼% interest rates.

Heck, now you're talkin' about 2%, 2 ½ %, maybe, 3% that you're saving yourself on your student loans.

Let's take our earlier example, when we're talking $250,000 in student-loan debt.

We're talking $5,000 a year! Maybe $7,500 a year that you're saving by refinancing with one of these companies.

You can finally get ahead on one these loans!

By the way, my friend and blogger extraordinaire PhysicianOnFire (www.physicianonfire.com) has an affiliate link where you can get $300 credit through SoFi. This is better than most other offers at $250.

You can find this exclusive deal at www.SoFi.com/PoF.

(FYI- An affiliate link means that there is some small compensation that the referring party receives for the referral. This is a win-win situation and I highly recommend doing so for those of you that refinance your loans)

THE "ORBITZ" OF STUDENT LOANS

Of all of these different experiences, the one thing I was thinking about the other day...

Wouldn't it be cool, if there were a company that was like Orbitz, where you could shop and compare all of these different companies at once?

Consider this... Given everything I've described so far, here's what happens:

You have to go, one-off, to EACH of these individual websites and submit your info. You have to submit your Social Security number, your date of birth, your driver's license information, your this and your that and your this and your, that TO EACH of these sites.

IT'S SOOOOO FRIGGIN' ANNOYING!

Yet, to save a dollar, I'm willing to put up with their annoyances, for the sake of my clients.

That kind of reminds me of Southwest. Southwest doesn't put their fares on Orbitz or Expedia, or on those other travel websites. You have to go to Southwest's own website in order to get a quote for an airline ticket.

Can't they just put it all in one doggone place??

Well, there's a new website I just found out about, called "Credible." (www.credible.com), where you can shop for your loans. As a matter of fact, GradSchoolLoans.com pointed out that site to me.

On www.credible.com, you can compare among seven to eight different companies.

(NOTE: Most of the companies we've referred to so far--SoFi, Laurel Road, etc.--are NOT currently on that site).

On credible.com, you can see if and at what rate a particular lender will refinance your debt if you are already a resident, or if you are in practice. **So, maybe you could swing an even better rate.**

To be honest? I haven't been on there yet more than a couple of times. Earnest & SoFi gave better offers in the limited cases I dealt with.

I'm looking forward to doing more of such website compare-and-contrast shopping in the future. Let me know your experiences. If you want to do it together, awesome! I'd love to see how that goes and how the experience compares to the one-offs.

Anyhow, there's a whole bunch of resources. I've done a whole bunch of different Podcasts, for most of these companies, whether it's SoFi or Laurel Road or LinkCapital. GradSchoolLoans.com and Common Bond are probably going to be discussed in future shows on my podcast- www.doctorfreedompodcast.

Now, I'd love to hear from you.

Are you considering PSLF? Are you considering refinancing your loans?

What does that look like? How are you making those decisions?

And what can I do to help provide you with the kind of scenarios that will help you understand which loan is best for you?

Which path is best for you?

ANOTHER WAY A PODIATRIST REFINANCED

In the spring of 2016, a podiatrist I worked with was so excited to transition to practice.

He did an extra year of fellowship (which was really unusual!) and he was stoked and ready to go. During the prior year, we had been discussing refinancing his debts.

However, he had no idea where he was going to land. He could have worked at a non-profit or at a for-profit.

He interviewed and interviewed, kept on looking, across the whole country. The place he landed was a private practice, a for-profit entity. There was no doubt in my mind that refinancing was definitely the right move NOW.

We explored a few of the companies together. It turns out that he was blessed by having family with a significant amount of money. He raised his head and came up with an idea. **What if he took a family loan at a 2.5% interest rate?**

I thought, that's a fantastic idea! He made it happen and is on track to be debt-free within the next few years.

RESOURCES MENTIONED:

SoFi:
www.SoFi.com/PoF

Podcast Interview with Dan Macklin:
www.doctorfreedompodcast.com/sofi

Laurel Road/ Darian Rowayton Bank:
www.laurelroad.com/student/loans

Podcast Interview with Alex Macielak:
www.doctorfreedompodcast.com/drb

LinkCapital:
www.linkcapital.com

GradSchoolLoans:
www.gradschoolloans.com

CommonBond:
www.commonbond.co

Credible:
www.credible.com

Earnest:
www.earnest.com

Physician On Fire Blog:
www.PhysicianOnFire.com

CHAPTER 11

HOW I BECAME THE DEBT FREE TERMINATOR BY 31
By Amanda Liu, MD

D o you have student loans?

Won't you love to destroy them and be debt-free?

What would you do with the cash flow you free up once your student loans are paid off?

I (Amanda) invite you to join me in a movement to terminate the deadly burden of student loans.

Under my cover as a mom, radiology resident, blogger, gourmet chef, & USMLE tutor, my true identity is a terminator, specifically programmed to terminate deadly student debts.

Below, I will share my weapons of termination in hopes of eliminating student debt on the scale of an entire generation.

STRATEGY# 1: INTENTIONALLY USING CREDIT CARDS

Yes, I said it. You SHOULD intentionally use credit cards to help you pay off your student debt much sooner.

Essentially, borrow at 0% to even negative interest rates to pay down your student loan faster.

This is a bit complex. So, please bear with me for a minute here.

- I charge all my expenses that are chargeable onto my credit cards and funnel my (limited) cash flow towards debts with interests. There are lots of variations in terms of what can be charged on a credit card.

- Some people's circumstances even allow them to pay for rent on credit card. At one point, I used to pay my landlord by charging her necessities such as gas and groceries on my credit cards. This takes a little more effort than just writing a check.

- Now, I buy thousands of dollars' worth of grocery gift cards (enough to last 6-12 months because once a year there's a 10% discount on gift cards). I also pay my electricity one year in advance. I talk more about this in my post *Discover Lent Me -31% Interest* mentioned in the resources at the end of this chapter.

- Funneling cash this way, often got me **negative** 1-5% interest, which gave me more cash to pay down student loans. But, unfortunately, there's a limit to this.

- This second method, balance transfer checks, usually allows for more aggressive paying down of a higher interest debt. You can discover how I did this at www.DrWiseMoney.com.

- The cheapest balance transfer checks I got was with Travelocity American Express at 1% transaction fee for 0% APR for a year. So by writing a check of $15,000 towards a debt such as student loan at @ 6.8% interest rate, I would save 5.8% for the next 12 months.

- The balance transfer transaction fee is charged up front, so just be sure that if your limit is $15,000, that you write a check in the amount lower than the limit enough to pay for the fee.

- This is to ensure that the check goes through, and you're not charged an additional fee. (I have never gotten a fee before, as I always err on the safe side.)

- There is one card that does *not* charge transaction fee for balance transfer if you use it within 60 days of account opening. *You can read about it on my website and checking the resources page at the end of this chapter.*

- Some banks allow you to open a new checking account by funding it with a credit card. You need to be very cautious with this. You need to make sure that **funding** is equivalent to a **purchase**, and not considered a **cash advance.**

- When your credit card company processes funding a new bank account as a purchase, that purchase will give you **cash back** (if your card offers cash back features). When your credit card company processes funding a new bank account as a cash advance, you will be charged an interest of 20-30% starting the day the transaction posts. So this method **only** works if your credit card company processes your act of funding a banking account as a *purchase.*

STRATEGY# 2: REFINANCE

As Dave mentioned in the last chapter, for a while residents and fellows had no refinancing options to lower their student loan interest. However, mid 2015, private banks began to offer student loan refinancing to residents and fellows so no one needs to suffer the 3-7 years of debt snowballing at 6.8+% during training.

The only drawback is that you forego loan forgiveness when you refinance. My mentor Dr. James Dahle at whitecoatinvestor.com commented that "student loan refinancing isn't new. It just went away for a few years. My class all refinanced at 1-2% back in 2003."

Here are some other alternative methods that you could consider:

Home Equity Loan. Should you be fortunate enough to own a home, home equity loan is frequently much cheaper than 6.8%.

It's a perfectly simple, passive, & effortless way to make your hard earned dollar go further. With a lower interest rate, every dollar you dedicate to your debt pays down a greater percentage of principle.

Plus, it's likely tax deductible as well!

Borrow Money From The IRS, Yes, you may be able to buy yourself some time by borrow from the IRS while being interest free.

If you do some locums and receive <u>1099s</u> both during training and as an attending, there is likely a big jump between your 1099 incomes during the transitional year.

Make sure to check out my post *Know Thy Worth and Be Thy Boss* on <u>www.drwisemoney.com</u> .

Since you pay taxes every quarter for your self-employment income (1099) based on prior year projections, you could seriously make a dent on your debt by delaying paying 1099 taxes for your first year out (higher 1099 income as attending) until the April tax filing deadline.

Effectively, you would have borrowed gobs of money from IRS interest free, with the very first penny made at the beginning of the full year as an attending riding 0% interest loan from the IRS for 16 months.

Since the IRS loves borrowing money from taxpayers (including you) interest free (every time you get a tax refund, you have lent the IRS interest free money), you can return the favor.

FINAL THOUGHTS

We terminators must unite. We ought to share ideas and weapons.

I invite all you soon-to-be terminators out there to work towards collaboration: it behooves us to help one another.

Don't we all wish we could be Jon Conner and send a debt terminator back in time?

Prevention is the best medicine. You want to destroy student loan Cyberdyne before Genesis.

RESOURCES MENTIONED:

Discover Lent Me Negative 31% Interest
http://drwisemoney.com/2016/03/09/discoverit-lent-me-3600-on-negative-31-interest-for-15-months/

How Balance Transfer Checks Work
http://drwisemoney.com/2015/03/11/how-credit-card-companies-compete-to-save-me-interest-on-student-loans/

The Credit Card I Recommended
http://drwisemoney.com/2015/04/15/these-credit-card-offers-do-exist-take-advantage-of-it/

Refinance Your Student Loans as Low as 1.95%
http://drwisemoney.com/2016/03/23/refinance-your-student-loans-to-1-95-fixed-interest-rate-start-funding-your-dreams/

Know Thy Worth and Be Thy Boss
http://drwisemoney.com/2016/05/24/know-what-youre-worth/

CHAPTER 12

5 REASONS WHY CREDIT CARD DEBT TRUMPS STUDENT LOAN DEBT

By Amanda Liu, MD

After reading the last chapter, you may be intrigued. On the other hand, maybe you are sitting here reading this and you are super skeptical.

You're wondering,... Really? Are you serious? She wants me to take out credit card debt?!?!?

Yes, I am serious! You can do this!

Are you kidding me? Look at the interest rate!

First off, the highest interest rate I have ever paid to a credit card company is 1.7%.

In fact, I have not paid a penny of interest other than that of my student loans (paying interest to department of education) since January 2014.

I have been indeed borrowing negative interest money to pay off my student loans and max out my retirement savings. Why would anyone choose 6.8% interest rate over negative to 1.7% interest rate debt?

REASON# 1: COMPARE FEDERAL LOANS VS 0% CREDIT CARDS

I know there's stigma and fear associated with credit card debt, but to me what's more scary is debt snowballing at 7% interest rate while I sleep, work, & eat...

FEDERAL STUDENT LOAN INTEREST RATES (FIXED)

	July 1, 2016 to June 30, 2017	July 1, 2015 to June 30, 2016
Undergrad Direct Stafford Loan – Subsidized	3.76%	4.29%
Undergrad Direct Stafford Loan - Unsubsidized	3.76%	4.29%
Graduate Direct Stafford Loan - Unsubsidized	5.31%	5.84%
Direct Parent PLUS Loan	6.31%	6.84%
Direct Graduate/ Professional PLUS Loan	6.31%	6.84%
Perkins Loan	5.00%	5.00%
HPSL (Health Professions Loan)	5.00%	5.00%

REASON# 2: THE ORIGINATION FEES

Check out the origination fees on the previous page! It's higher than credit card balance transaction fees.

Now, the balance transfer offers I had gotten throughout medical school included 1% transaction fee for 15 months balance transfer checks at 0% APR.

To put in plain language, private banks were charging me 1% up front for me to borrow money up to my credit limit for 15 months interest-free.

So what is the true cost/effective interest of such a debt/offer? 1% divided by 1.25 years = 0.8%. You can see it 2 different ways to compare apple to apple.

1. The balance transfer has $0 origination fee and effective 0.8% interest annually vs. student loan with 4.27% transaction fee (which becomes your principle the moment your loan disburses) & 6.8% interest rate.

2. The balance transfer has 0.8% origination fee and effective 0% interest annually vs. student loan with 4.27% transaction fee (which becomes your principle the moment your loan disburses) & 6.8% interest rate.

No matter which way you look at it... how are student loans a better deal than credit card debt?

Nowadays, my balance transfer offers are either 0% transaction fee for 15 months of 0% interest rate, or 2% transaction fee for 14 months of 0% interest rate.

Still beats the federal student loans!

FEDERAL STUDENT LOAN ORIGINATION FEES

	October 1, 2015 - September 30, 2016	October 1, 2014 - September 30, 2015
Undergrad Direct Stafford Loan - Subsidized	1.068%	1.073%
Undergrad Direct Stafford Loan - Unsubsidized	1.068%	1.073
Graduate Direct Stafford Loan - Unsubsidized	1.068%	1.073%
Direct Parent PLUS Loan	4.272%	4.292%
Direct Graduate/ Professional PLUS Loan	4.272%	4.292%
Perkins Loan	0.00%	0.00%
HPSL (Health Professions Loan)	0.00%	0.00%

THE REASON# 3: THE REWARDS/INCENTIVES TO CHARGE ON CREDIT THAN TO ASK UNCLE SAM

Now, let's talk about borrowing negative interest money from credit cards to fund your education rather than paying 4.3% loan origination fee with 6.8% interest accrual on what you borrowed plus the origination fee.

In medical school when I charge my trimester tuition of $15k every four months, I make anywhere between $150-300 cash back. I could make even more if I redeem the points for gift cards or flight mileage rather than cold cash.

If I were to borrow from Uncle Sam the same $15k for four months of medical school education, I would have been charged a $600 origination fee, and have a debt principle of $15,600 snowballing at 6.8% interest rate the minute the loan disburses. In comparison, it is a few days later when I get the check and cash it with cards.

That's a $900 difference up front with cards. Then either I ride the 0% interest on credit cards for 18 months, or I let the 6.8% interest from Uncle Sam crush me for the same 1.5 years.

What would you choose?

THE REASON# 4: BANKS COMPETE, YOU WIN

Discover, Bank of America, Wells Fargo, Chase, Citibank, & American Express, were competing for my debt. They were hoping to bait and switch on me (i.e. bait me with introductory promotional interest rate of 0% then switching to 17% after promotion ends.) When banks competes, borrowers win.

Do you know how many competitors are against Uncle Sam? Nada, zero, zilch.

As Uncle Sam monopolizes the "federal" student loans market, they charge whatever they like. While private banks can borrow 0% interest rate (prime rate for a while as you recall) from feds & taxpayer dollars from you and me.

The education arm of the Feds decided arbitrarily to charge those who want to advance their education 6.8%. You see where our national value lies. The value is clearly in business, not in education.

Because of all the banks competing to bait and switch on me, I never ran out of offer. As I write right now, I have $0 student loans, $0 consumer/credit card debt, and the only debt I will have in a few weeks is the mortgage of my 2nd/dream home. I still get 10+ credit card balance transfer offers, as the banks hope that I will bite their bait and stick around for the switch.

While these banks are the same faceless corporations who charged my dad 30% interest rate when he missed a payment 10 years ago, I don't feel bad to fund my education with their *negative* to 1.7% interest rate loans.

THE REASON# 5: WHAT TO DO WHEN LIFE GOES DOWN THE TOILET

When things goes really really badly, student loans stay, credit card debt is discharged in bankruptcy.

What's the message here? Don't mess with Uncle Sam. While I paid back every penny I owe (principle + interest) to Uncle Sam and private banks.

White Coat Investor raised a good point… *"I even thought to myself, well, what if you just left all that debt on the credit cards and just declared bankruptcy at the end of medical school?*

Student loans don't go away in bankruptcy, but credit card debt sure does. By the time you get out of residency 3-5 years later, that bankruptcy is almost off your record. Unethical? Of course. But geez, I can't say it wouldn't be tempting when staring a $400K student loan in the face."

While per my personal moral standards, I will not advise anyone to do the above. I'm sure that WCI also did not intend to encourage intentional bad debt.

What I do see though if a PGY3 gets disabled and no longer can finish medical training or practice medicine at all, yet he/she has 400k of student loans at 7% interest rate from the feds.

He indeed would have been better off to have these debts on credit cards and file bankruptcy.

FINAL THOUGHTS

Credit card debt has been known to negatively shape a financial future. 20%+ in interest is no way to finance a college education, especially when that interest compounds.

However, when you turn the tables on the banks and the credit card companies, and pay these things off quickly- you can be wayyyy ahead of the game.

Imagine what it would be like to pay 0% in interest INSTEAD of 6.8% with origination fees.

What do you think? Are you ready to make the leap?

In short, I'm way more scared of borrowing from Uncle Sam, the monopolizer of government issued student loans, than borrowing from credit card companies/private banks in a highly competitive market, favoring consumers.

How about you? How do you feel about these strategies?

Let me know. You can find my contact info & more of my adventures and advice at www.DrWiseMoney.com.

CHAPTER 13

BIG BAD BANK LESSON #1: IT TAKES NOTHING TO MAKE MONEY.

By Amanda Liu, MD

People say it takes money to make money. Not true. Just look at the big bad banks. It takes big banks *Nothing* to make money. What do I mean?

Theoretically, anyone, without any assets, can make an infinite amount of money if he or she is a bank. A bank in the US can borrow money from the tax-payers (federal government) at 0% (currently 0.25%) interest rate, then lend this money out to consumers at rates ranging from 2% to 30+%.

Isn't that incredible?

Hypothetically, a penniless bank, borrows money which it in turn lends out at pretty much any rate it chooses. The only risk here is default, but when you have 30% growth off of borrowed money, you can withstand some defaults.

Banks don't have their own money… they don't even hold onto the money they borrow.

All banks do is to create this money current, making themselves the conduit that the money river flows through, they take interest difference (between the lower rate they borrow at and the higher rate they lend at), transaction fees, loan origination fees, and other fees galore.

Truth be told- they do have capital requirements, but there's a point to be made here.

Why don't we 'every-day people' also turn ourselves into mini banks?

Learn from the big bad banks, and make ourselves a channel for money river…

Here's how I did it and made $3,300 by a few mouse click using about 2-3 hours of my time.

Now, there is online sources that states that the Citibank is no longer taking credit card funding for new Citibank bank accounts.

Since it is not an official statement from Citibank and you'd still like to try, go ahead… but don't get mad if it doesn't pan out. Consider it a lesson.

The point here is learning from the big bad banks and seizing the opportunity to be the bank yourself for once!

Here's the synopsis of what I did to create a money river with me being the conduit that directs the money flow and profits off of it.

1. Open Citibank online account
2. Fund the Citibank account with my BOA Visa Cash Reward card
3. Once money is in the new Citibank account, pay off BOA credit card
4. BOA credit card gives me 1.1% cash reward
5. Since I funded $50k, I get $550 cash rewards in my BOA checking account
6. Close the Citibank bank account after 30 days (not much longer, I wanted to avoid the monthly fee).
7. Open a new Citibank bank account online and then repeat 1-6 steps. Every cycle, I make $550.

So I made $3,300 when I had $0 in my bank accounts. It was just 1.1% cash back.

Just imagine 2% cash back, 3% cash back, or even better 30% cash back (which is the kind of deal our big bad banks are getting.)

HINT: You have to be on the ball. Remember to be VERY aware of the fees that the banks charge. You have to play their shell game and move the money around before they charge their fee. Make sure to set calendar reminders a few days before the deadline.

Big bad banks did nothing but moved tax-payer's money around… the more profitable directions the big bad banks manufacture, the more rewards they make for themselves. (16% interest rate on credit cards, 30% once credit cards defaulted, 5% on student loans, 4% on mortgage, 6% on used car loan, etc.)

In summary, it does not take money to make money. Money can be collected from the money river flowing through your property. Much like all living things on the banks of river thrive from what the river offers as it runs by.

Be the bank, capture what the cash river offers you as it goes by.

Let me know what you think of this advice after you try it out.

What did you find in your experiment?

You can find my contact info & more of my adventures and advice at www.DrWiseMoney.com.

CHAPTER 14

YOUR ACTIONGUIDE
TO BEING DEBT-FREE

By Dave Denniston, CFA

As a physician, you've made a commitment to helping others and your community.

Now make a plan to pay off your debt!

Consider for a moment: Could you utilize one of the programs we have discussed?

Could you utilize the Big Bad Banks money like Dr. Liu has brilliantly illustrated for us?

Could you borrow at 0% for the next 12 to 15 months and then find another card *or simply pay that debt off?*

Could you refinance your loans with an outside firm like SoFi, DRB, or Earnest?

Also, one other topic that isn't discussed enough: What if you could COMBINE two debt forgiveness programs simultaneously?

For example, you could enroll in PSLF, work for a non-profit in an under-served area, and then at the SAME TIME, do a state forgiveness program for 2 or 3 or 4 years (whatever the minimum commitment is).

This could hedge the bet of the federal government taking the punch bowl away from the party. This way, you have substantially less debt no matter what happens.

If, as a young physician, you focus on paying off your debts, if you save for a rainy day, if you live within your means and put money away for retirement, then you can do the things you've long dreamed of doing and you'll be well down the road to financial independence.

Take the next step and complete the Action Steps on the next few pages!

❖ **Action Step:** Gather together the data that you will need in order to assess all of your liabilities.

We'll cover how to prioritize liabilities in the next step, but for right now, just collect the information.

Below we have included an example for your review.

Liability Description	Company	Principal Owed	Interest Rate	Maturity Date	House, Consumer, or Business?	Fixed or Variable?
Home Mortgage	Bank of America	$ 250,000	4.50%	1/1/2032	House	Fixed
HELOC	Bank of America	$ 10,000	2.50%	n/a	House	Variable

Liability Description	Company	Principal Owed	Interest Rate	Maturity Date	House, Consumer, or Business?	Fixed or Variable?
Student Loan	Sallie Mae	$ 150,000	6.80%	1/1/2033	Consumer	Fixed
Car Loan	BECU	$ 10,000	5.00%	1/1/2017	Consumer	Fixed
Credit Card	Bank of America	$ 30,000	11.00%	n/a	Business	Variable

Note how nearly every spot has something in it.

I entered in "n/a" for maturity date for the lines of credit, since there is no fixed maturity date.

One of our goals will be to create a specific maturity date, once we have identified the loans that are a priority.

If you are confused by this process, or if you need help getting the information together, feel free to give me a call or send me an e-mail.

Now that you have defined your reality, take some time to reflect on how you've gotten to this point.

- ❖ **Action Step:** What positive things have I done to put myself in a <u>healthy</u> financial position?

- ❖ **Action Step:** What are some areas where I have sabotaged myself? <u>Where can I improve?</u>

- ❖ **Action Step:** Who are some positive mentors/role models who can help hold me accountable to my goal of becoming debt-free? Speak with that mentor and make a schedule of meeting with them at least twice a year, if not quarterly.

- ❖ **Action Step:** Review ALL of the debt strategies listed throughout the last few chapters. Could you utilize a debt forgiveness program?

Is PSLF a good fit?

Are you married? Should you consider a strategy such as filing taxes separately, even if you are part of a married couple?

Should you consider a state-sponsored program?

Could you COMBINE PSLF with a state-sponsored program, in order to hedge your bets?

If you are not planning to work for a non-profit, could you refinance your debt through a company like SoFi or DRB?

Are you willing to make a commitment to working in the for-profit arena as a resident?

Could you refinance your debt while you are still in residency?

Now that you are armed with all the tools- it's implementation that allows you to achieve results—just as merely *thinking* about exercise won't make you physically fit.

Write down your commitment to implement your plan and what you will do to stay on track. Show your plan to your mentor. Brainstorm with them: What can you do to avoid financial distractions and pitfalls?

FIRESIDE CHAT WITH DR. ERIC GANTWERKER

In the first fireside chat I ever held, we discovered that this wonderful and caring physician took 18 years after graduating high school to transition to practice.

As a matter of fact, he moved 18 times in 18 years as he went through this journey.

Throughout this time, debt has been a chain around his ankles. During the chat, he mentioned that one of his greatest mistakes was not applying for PSLF earlier. He encouraged residents to apply as soon as possible so that they can get the clock ticking on their potential debt forgiveness.

He also shared the best decision he's made since transitioning to practice. HINT: It's not the debt he's paid back.

Learn more about Eric's journey at:
www.doctorfreedompodcast.com/eric

RESOURCES MENTIONED

Fireside Chat with Dr. Eric Gantwerker
www.doctorfreedompodcast.com/eric

Details on PSLF:
http://studentaid.ed.gov/repay-loans/forgiveness-cancellation/charts/public-service

State-Sponsored Debt Forgiveness Programs:
https://services.aamc.org/fed_loan_pub/.

National Health Service Corps Repayment Programs:
http://nhsc.hrsa.gov/loanrepayment/

Your Insurance Guide

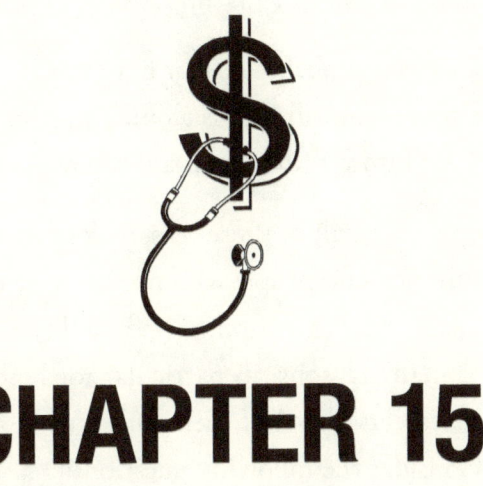

CHAPTER 15

THE DISABILITY INSURANCE GUIDE FOR PHYSICIANS
By Dave Denniston, CFA

Perhaps you are a young physician having just come out of medical school. No more tests or studying; you've just landed your feet!

Or maybe you are now in your residency or your fellowship and making a little bit of dough. You're starting to think about the future.

You are now being hit up by salesperson after salesperson who wants to sell you all kinds of financial advice--particularly advice on insurance. Maybe you've received a seminar invitation or two (or three or four or five)?

They are hitting you up in rounds. It may even feel like they are breathing down your neck and beating down your doors.

Why? Because you probably do not have much in actual cash savings yet, and selling you insurance is a quick way for a smart salesman to make money for himself (the salesman) today.

However, even given all of those reasons for you to be a little wary of the many insurance salespeople who may now be contacting you, it is important to note that life insurance & disability insurance can both be necessary for many reasons, in particular for high-earning doctors with their whole careers ahead of them and often a great deal of debt behind them. Consider the following questions:

How would your family manage without your income if you died or become disabled?

How much do you have in reserve for the "stuff happens" in life events?

The bottom line: insurance, while necessary to protect our families, is a cost that should be minimized.

Explore and learn with me how you can minimize those insurance costs while maintaining an appropriate amount of insurance for your specific situation.

NOT ANOTHER PLUG FOR AN INSURANCE COMPANY

First of all, this is not another plug for any specific insurance company. I am an entrepreneur. I am not affiliated with any insurance company and I am doing this to help educate you and your family on how to navigate the confusing number of choices that are available to you.

Regardless of whom you work with, make sure that you select someone who offers you a multitude of choices and who educates you on the cheapest option as well as the most expensive option. They should empower you with the information necessary for you to make an educated decision regarding either life insurance or disability income insurance.

Frankly, insurance of all types is over-sold, in my opinion. My basic philosophy is that it is when you are most vulnerable that is when you most need insurance. <u>I encourage you to save and save and save, so that you don't need an insurance company anymore and you can be "self-insured" down the road.</u>

Ultimately, my job is all about helping you to make smart choices about your money, and I believe that this chapter will be a standard against which to measure others as you speak with insurance consultants.

In the next few chapters, we are going to address the following:

➤ Disability Income Insurance: Overview. Group versus Individual Benefits
➤ Disability Income Insurance: Managing Costs Plus The Bells & Whistles (Riders)
➤ Disability Income Insurance: How to Compare One Company versus Another
➤ Life Insurance: Overview. How Much should I Have?
➤ Life Insurance: Term versus Cash-Value Insurance

The exercises in these chapters are incredibly important in finding the right insurance for you. For many of us, it can be difficult to look at the issue of insurance. Sometimes, we need a guiding hand.

<u>Don't hesitate to ask for help if you find that you are procrastinating or that you just can't stand looking at the data on your own. You can contact me at dave@doctorfreedompodcast.com.</u>

DISABILITY INCOME INSURANCE OVERVIEW

First, let's understand disability income (DI) insurance.

Why would you want DI insurance?

According to the Council of Disability Awareness and the Social Security Administration, illnesses and injuries such as cancer, heart disease (heart attacks), diabetes, back pain, physical injury, and arthritis are common causes of both long-term and short-term disabilities.

Further, almost 3 in 10 of today's 20-year-olds will become disabled before reaching age 67. That's 30 percent!

Additionally, the *American Journal of Medicine* (Vol. 122, No. 8) states that every 90 seconds someone files for bankruptcy in the wake of a serious illness.

Disability and the financial crises it can create are a very real and present problem. According to the Standard Insurance Company, when disabilities do set in, the average duration of disability for those affected who are below age 50 is five to six years.

Ok... Let's take a breather for a sec...

....

......

....

Blah, blah, blah... Your eyes are rolling to the back of your head. Kind of boring stuff, right?

THE NO-B.S. TRUTH

I know what you are thinking: "Okay, I get it. People get injured. I don't want to lose income. I need DI. Really, is that all you got?"

All right, listen.

This is real-life stuff. Seriously.

This kills many physicians' financial futures.

But you know what? I have to be honest. This isn't for everyone.

In my opinion, if you aren't doing surgeries or procedures…

If you could do your job in a wheelchair and AREN'T concerned with your dexterity or your ability to use your hands, and if lack of dexterity wouldn't derail you professionally…

You know what?

<u>Maybe DI isn't for you.</u>

<u>I firmly believe that DI actually ISN'T necessary for all physicians.</u>

Whew, there, I said it…

Glad to get this off my chest.

However, IF YOU ARE DOING SURGERIES OR PROCEDURES…

IF DEXTERITY AND THE ABILITY TO USE YOUR HANDS is important to you and to your career…

Consider this question:

How would your financial situation change if you weren't able to perform at your current capacity for five or six years?

What would be the impact on your retirement?

How would a disability affect your college-education planning for your kids?

What possible problems could this create for you?

Voilà! Disability Income (DI) insurance was created to protect against this risk.

There are two basic types of disability insurance: group DI and individual DI.

➢ Group DI insurance is an employer-provided benefit.

This type of DI may cost you little to nothing. Typically, the employer is using their own *pre-tax dollars* to provide a benefit to you, the employee.

They are utilizing the law of large numbers in order to minimize the cost. Using data on the age of employees, employee compensation, the employer's claims history, and other information, the insurance company will work with the employer to determine the cost to provide a DI benefit for all employees.

Note that because the benefits are funded with pre-tax dollars (by the employer, usually) the future benefit will be taxable income to you.

As long as you are using a certain level of benefit within the group, you don't have to go through any individual medical underwriting.

This means that they don't look at your height, weight, or medical history. They won't take a blood draw or other bodily measurements.

For someone who is in generally poor health, who has an above-average build, or who has a negative medical history, maximizing group insurance is a wonderful way to obtain insurance when you could not normally afford individual insurance.

Whereas...

> ➤ <u>Individual DI insurance</u> will have the insurance company measuring many different criteria in order to determine whether you are healthy enough for the company to offer you a policy; or, even if you are offered a policy, the company may possibly add on some exclusions for particular pre-existing health conditions.

In contrast to the situation with group DI, for an individually-purchased disability insurance policy, you will be spending (unless you are self-employed) after-tax dollars from your own personal bank account. Note that because the benefits were originally funded with after-tax dollars (funded by you, rather than by an employer), the future benefits will NOT be taxable.

Besides the tax advantage just noted, *<u>the major benefit of individual DI is that you have the ability to customize a policy specifically for your needs.</u>*

For example:

Would you like your policy benefit to last until age 65 or for just 5 years? It's your choice!

Would you like your policy to have a waiting period of 90 days or 180 days? It's your choice!

Would you like to have the ability to increase (or decrease) your coverage? You can do it.

Would you like to cover a small part, a medium-sized part, or a large part of your income? It's up to you!

We'll discuss in more detail the cost trade-offs in dollars for each of these components (as well as others), as well as the differences between the "Big 6" insurance companies.

There are six traditional companies (the "Big 6") that insurance brokerage companies typically consider for physicians: Standard, Principal, MetLife, Union Central (First Ameritas), Berkshire (Guardian), and MassMutual.

This is because each of these companies has included (or offers as an additional rider) a benefit of "Your Occupation" or "Own Occupation." We'll discuss this in more detail later.

THE ANESTHESIOLOGIST WHO BECAME DISABLED

One of my clients is an anesthesiologist. She and her husband immediately purchased DI during her first year of residency. As a matter of fact, they were so eager to do so that I was a bit hesitant.

Living on the budget of a resident, they were scraping by and had a little bit of credit-card debt.

I swallowed in uncertainty. I wasn't really sure that this was the right thing to do.

Why were they so set on getting DI?

Her father-in-law was an anesthesiologist and he had, unfortunately, become fully disabled in his mid-40's. To be honest, I can't remember why. It may have been an accident or a disease. I'm not sure. However, for them, they'd seen it first-hand: a physician who had to stop practicing medicine due to unforeseen circumstances.

Here are some general guiding thoughts that I would like for to you consider:

➤ First, how dependent is your family on your income?

➤ Are you the sole breadwinner or is yours a two-income family? If yours is a two-income family, could your spouse's income cover living expenses and meet your debt obligations?

➤ The more your family is dependent on your income, the more you will need disability income insurance.

➤ While you have a substantial student debt load/mortgage, you should strongly consider having an individual DI policy.

Note that new rules released by the U.S. Department of Education could affect many Social Security beneficiaries who have student loan debt. Effective July of 2013, the repayment obligations of student loan borrowers who are deemed by Social Security to be permanently and totally disabled may be discharged.

The key here is "permanent," and you require approval from the Social Security Office. If you are partially disabled or if you are disabled for a period that is not "permanent," you will still owe regular payments on your student debt.

➤ The longer the period between your initial disability and your retirement, the greater the potential income and savings for you to lose.

<u>After you have been out of residency/fellowship for 15 to 20 years, you should hopefully be close to "self-insured," and your need for career-long disability insurance should drop.</u>

At that point, consider a cheaper individual policy (i.e., a policy offering lower benefits or benefits over a decreased period of time) or just stick with the group term benefit.

➤ Make sure that, in addition to investing in 401k/tax deferred savings vehicles, you are consistently saving in a non-qualified account (an account not subject to early-withdrawal penalties: for example, a bank account or a "regular" investment account). This type of savings will help to protect you and your family against "rainy-day" conditions such as a disability.

Once you have saved $150k or more (in today's dollars) in a combination of the 401k and non-qualified accounts, I would consider you to be at least somewhat partially "self-insured," assuming that your annual living expenses are $60k or less (in today's dollars).

The more savings that you have in non-qualified accounts, in particular, the better. Although it should be remembered that in most cases you can take a tax-free loan on a 401k of up to 50% of its value, or $50,000 (whichever is less), as long as you pay back the loan over a handful of years.

Below is a summary chart that is NOT comprehensive, but that explains the basics of comparing and contrasting a typical group DI policy versus an individual policy.

Typical Group Policy	Typical "Big 6" Individual Policy
Little to no medical underwriting required. Neither age nor health will usually matter at time of application.	Substantial medical and financial underwriting required. Better the younger and healthier you are.
Cheapest option.	Usually more expensive, depending upon the riders you select.
If paying for the policy yourself, you get a tax deduction. If benefits are paid by another, they are taxable.	Benefits are tax-free (assuming paid from post-tax account).
Premiums paid monthly at no extra cost.	Premiums best paid annually. Quarterly or Monthly costs more.

Typical Group Policy	Typical "Big 6" Individual Policy
After 24 months, any new income-producing job (i.e.,NOT "Own Occupation,") will from benefit.	Comes with or can add rider for "Own Occupation." This means that you can work anywhere after being disabled and can maintain DI benefit *plus* new earned income.
Contract terms, premiums, and benefits can change at any time.	Comes with or you add a rider for "non-cancellable & guaranteed" to ensure that premiums and benefits cannot change.
Policy ends when you end employment.	Policy is carried by you, before and after employment with any employer.
No partial-disability benefit.	Has a partial disability benefit (15% to 20% or more loss of income minimum).

> ➢ Make sure to have the agent who is helping you obtain quotes from three different insurance companies and explain the differences between them. Don't be shy to ask how much business the agent does with each of these companies and how he or she decides between them.

❖ **Action Step:** Gather together information on your current group disability income policy and fill in the information below:

Current benefit amount:_____

_____

What is the maximum benefit?_____

_____

How long does it cover "Own Occupation" ?_____

_____

How much does it cost me?_____

_____

Does it include partial disability?_____

_____

How (if at all) does it rise with inflation?_____

_____

❖ **Action Step:** Consider your current financial situation. How much debt (consumer, student, home) do you have?

How much in liquid accounts (bank, investments) do you have?

Do you have enough in liquid assets to cover 5 to 10 years of living and liability expenses?

Would your current group term DI cover those needs? For how long?

Are you a sole breadwinner or does your family have two incomes?

Write down below your thoughts on the need for individual DI for your situation._____

_____

_____

CHAPTER 16

14 FACTS ABOUT DI THAT WILL IMPRESS YOUR FRIENDS

By Dave Denniston, CFA

MANAGING COSTS, PLUS BELLS & WHISTLES

Now that we've covered the basics of DI policies and why they are necessary, let's explore the costs for given benefit levels.

First, depending upon your specialty and area of practice, the costs can vary. The insurance companies can deem one specialty as "less risky" than another.

The "less risky" specialties are cheaper to insure! According to the Disability Insurance Website, a common insurance carrier listed 5 different rankings (least to highest risk): Class 4m, Class 3m, Class 2m, Class 1m, and Not Eligible. Below is a table of how common medical occupation specialties are classified:

Class 4m		Class 3m	
Audiologists	Neurologists (no surgery)	Anesthesiologists	Surgeons (all specialties)
Cardiologists	Internists	Dentists	Podiatrists
Dermatologists	Ophthalmologists	ER Physicians	
Family Practice	Pathologists		
Radiologists	Psychiatrists		

Let's introduce several terms that determine the cost of the policy:

1. **Monthly Benefit**: This is the monthly paycheck you would receive from an insurance company in lieu of income produced by the normal practice of your profession (assuming total disability or assuming a percentage payment, in the case of a partial disability). The higher the benefit to be paid in case of a disability, the higher the premium cost for the policy. The lower the anticipated benefit, the lower the premium cost.

2. **Benefit Period**: This is the maximum length of time that your benefit can last. This could be 2 years, 5 years, 10 years, "up to age 65," or "to age 67." The longer the period covered by the DI policy, the higher the premium cost. The shorter the period, the lower the cost.

3. **Waiting Period**: This is the amount of time between when your disability claim is accepted and when your monthly benefit starts. This is a little different. The longer the insurance company goes before it has to begin paying your claim, the lower the premium they will give you. The longer the period, the lower the cost. The shorter the period, the higher the cost.

PODIATRISTS: WATCH OUT!

Earlier this year, I saw a podiatrist client who was transitioning from a residency in a hospital to a fellowship in private practice.

I had originally told him that we should wait for a couple of months, until he transitioned to practice, before we shopped for individual disability coverage.

I usually suggest this for a few reasons:

#1: Cash Flow Is Relatively Minimal in Residency!

#2: Most residents/fellows want to save for their first home, and the cost of a DI policy takes away from that ability.

#3: Most are relatively young, extremely healthy, and easy to underwrite.

However, his private practice employer didn't offer group DI !!!

We had to go shopping earlier than I normally like to do. Here's what happened:

We couldn't find a single insurance company that would offer a benefit of more than $2,000/month while he was in residency or fellowship!!

I had never run into this issue before.

Apparently, insurance companies are particularly averse to podiatrists and only take them on ONCE THEY ARE ATTENDING PHYSICIANS (or have signed a contract to be an attending).

The lesson here: Be aware of your specialty and of any restrictions!

What should you ask an agent for?

I generally suggest that you ask for information (we call this a quote in the insurance world) on a couple of different monthly benefits levels (to replace 40%, 50%, or 60% of your income if you're in practice; $40k, $50k, or $60k if a resident/fellow), as well as for information on a couple of different benefit periods.

The cheapest possible individual policy you could purchase is one with an extremely low monthly benefit (e.g., $1,000/month), a 2-year benefit period, and a 365-day waiting period.

By contrast, **the most expensive individual policy** you could purchase is one with an extremely high monthly benefit (think 80% of your compensation while in regular practice), a benefit period lasting to age 67, and a 60-day waiting period.

For most residents, fellows, and young practicing physicians who have a great deal of student debt and very little in liquid assets, I suggest a medium monthly benefit (40% to 60% of practicing income), benefits to age 65, and a 90-day waiting period as being fairly reasonable.

As we mentioned earlier, <u>the more you become self-insured, the less DI benefits you need to purchase from an outside insurer.</u>

Let's say that your student debt is now paid off and that you have built up a portfolio of $500,000 (in today's dollars) & your living expenses are one tenth of that. It's worth taking a look at your current policy and changing the benefits accordingly by lowering the monthly benefit, decreasing the benefit period, and increasing the waiting period. You don't need as much insurance now! Why pay for something you don't need?

Let's explore a few examples:

Below is a table taken from an illustration for the Standard for an Anesthesiologist, **age 26**, at $3,000/month benefit. The costs listed below are annual premiums.

Benefit Period	Waiting Period			
	60 Days	90 Days	180 Days	365 Days
2 Years	$1,054.10	$785.90	$716.18	$627.87
5 Years	$1,244.51	$936.04	$854.72	$749.82
10 Years	$1,524.10	$1,143.91	$1,045.36	$916.99
To Age 65	$2,143.71	$1,597.76	$1,459.72	$1,281.74
To Age 67	$2,260.27	$1,683.65	$1,538.61	$1,350.26

Below is a table taken from an illustration for the Standard for an Anesthesiologist, **age 26**, at $3,750/month benefit. The costs listed below are annual premiums.

Benefit Period	Waiting Period			
	60 Days	90 Days	180 Days	365 Days
2 Years	$1,234.25	$915.12	$834.61	$732.80
5 Years	$1,445.97	$1,079.14	$985.67	$865.85
10 Years	$1,774.23	$1,322.19	$1,208.34	$1,061.22
To Age 65	$2,511.59	$1,861.76	$1,701.15	$1,495.19
To Age 67	$2,649.60	$1,963.17	$1,794.21	$1,576.24

Below is a table taken from an illustration for Principal for an Anesthesiologist, **age 26**, at $3,750/month benefit. The costs listed below are annual premiums.

Benefit Period/ Your Occupation Period	Disability Base Elimination Periods (In Days)				
	30	60	90	180	365
To Age 70/To Age 70	$3,728.72	$2,200.52	$1,803.29	$1,680.09	$1,571.76
To Age 67/To Age 67	$3,654.80	$2,126.94	$1,730.04	$1,606.86	$1,498.52
To Age 65/To Age 65	$3,600.80	$2,072.27	$1,675.36	$1,552.84	$1,444.17
5 Year/5 Year	$2,323.48	$1,209.48	$893.49	$815.11	$731.76
2 Year/2 Year	$1,994.65	$966.15	$653.72	$534.02	N/A

Below is a table taken from an illustration for Principal for an Anesthesiologist, **age 29**, at $3,750/month benefit. The costs listed below are annual premiums.

Benefit Period/ Your Occupation Period	Disability Base Elimination Periods (In Days)				
	30	60	90	180	365
To Age 70/To Age 70	$4,144.18	$2,637.23	$2,162.38	$1,990.25	$1,855.25
To Age 67/To Age 67	$4,048.33	$2,542.40	$2,066.87	$1,895.76	$1,760.09
To Age 65/To Age 65	$3,977.45	$2,471.19	$1,995.98	$1,824.54	$1,688.87
5 Year/5 Year	$2,534.01	$1,437.84	$1,052.37	$951.21	$852.16
2 Year/2 Year	$2,125.75	$1,133.95	$755.25	$813.47	N/A

Below is a table taken from an illustration for Principal for an Anesthesiologist, **age 26**, at $5,000/month benefit. The costs listed below are annual premiums.

Benefit Period/ Your Occupation Period	Disability Base Elimination Periods (In Days)				
	30	60	90	180	365
To Age 70/To Age 70	$4,740.50	$2,817.00	$2,313.50	$2,154.50	$2,015.50
To Age 67/To Age 67	$4,646.55	$2,723.00	$2,220.00	$2,061.00	$1,922.00
To Age 65/To Age 65	$4,576.50	$2,652.50	$2,149.50	$1,991.50	$1,852.00
5 Year/5 Year	$3,261.00	$1,697.50	$1,254.00	$1,144.00	$1,027.00
2 Year/2 Year	$2,799.50	$1,356.00	$917.50	$749.50	N/A

Let's review some of these differences.

Rates Change. First, I have used these illustrations as snapshots in time. The insurance companies can change their rates at any time and these examples may or may not reflect your own situation, which is dependent upon your specialty, your age, your health, etc.

We can see from each of these illustrations that there are substantial differences in premium if you move to the extremes.

The Difference. For example, if you use a 90-day elimination period and a benefit period to age 65 for the $5,000 illustration above, you are right in the middle at $2,150/year.

If you cut the elimination period to 30 days, you more than double the cost, at about $4,600/year!

Whereas if you cut the benefit to only 2 years, you cut the cost by more than one-half, to about $900/year!

Note that there is <u>some difference</u> in premium between benefit periods to age 65 and to age 70, or between 90-day and 180-day elimination periods.

However, the <u>dramatic difference</u> occurs when you cut the benefit to 5 years or less or when you increase the elimination period from 30 days to 60 days at Principal (or increase the period from 60 days to 90 days at the Standard).

Additionally, there is a dramatic difference in premium when you wait to receive disability income. Using the $3,750 monthly benefit, 90-day elimination period, benefit to age 65: At age 26, the cost is $1,675.35/year whereas at age 29, the cost jumps up dramatically, by almost 20%, to $1,995.98/year.

<u>In this example, if you decide to buy the DI insurance as a resident</u>, you are giving up the use of 3 years of premiums ($5,026.05) in order to avoid a higher annual cost ($320.63/year).

This is about a 15-year break-even. If you hold this DI policy past age 41 (assuming you purchase the policy at age 26), you will have made the best possible financial decision.

I think either is a fine decision (buying at age 26 or buying at age 29); the main point is to have the protection in place.

However, keep in mind that <u>you want to lock in your rates while you are a resident or fellow</u>. I usually suggest to pick up insurance in your final year of residency in order to not have financial strain.

Most of the "Big 6" insurance companies will allow you to underwrite up to a $5,000/month benefit or $6,000/month benefit before you enter practice. If you are in primary care, this will ensure that you are cover close to 40% to 60% of your income covered, between your individual DI and group DI.

As a specialist, you may need an additional policy after you transition to practice.

Once you are in practice, the insurance companies typically restrict how much they will underwrite in coordination with your group policy, which is far more limited in benefits and in your ability to use it.

A TALE OF TWO CITIES OF POOR HEALTH & WHY INDIVIDUAL COVERAGE ISN'T ALWAYS THE ANSWER

While our focus here so far has been on getting individual coverage through an agent, sometimes that isn't always the answer.

A few years ago, a physician with a specialization in internal medicine had just signed his first contract. He was so excited! He was deeply, deeply in debt--well over $250,000--plus he was helping to pay off family loans that totaled more than another $150,000, and in addition to all that, he was helping to financially support his parents.

He was scared, frustrated, and looking for direction.

In my opinion, he definitely had to get individual coverage. He was single and his extended family was dependent on his income.

However, his health was poor! He was 150 pounds over his ideal weight. He had issues with depression, asthma, heart palpitations, and much more!

With all of this going on, his medical records were thicker than a textbook!

I wasn't too keen on the individual coverage. I didn't think he would have much of a chance of qualifying for it, but decided to try anyhow.

Yeah, we failed. Couldn't get coverage. Ain't gonna happen.

However, that's not the end of the story. There's more...

A TALE OF TWO CITIES OF POOR HEALTH & WHY INDIVIDUAL COVERAGE ISN'T ALWAYS THE ANSWER (CONTINUED)

His current employer in residency allowed him to CONVERT his existing group policy <u>WITHOUT UNDERWRITING</u>.

The agent from the employer tried to sell him a policy with lots of bells and whistles. I advised him to strip it down, but still maintain great coverage! That was awesome!!!!

<u>If your health is less than optimal, make sure to check your residency program's group coverage, to see if you can convert the group policy to an individual one.</u>

<u>Fast forward to one year later</u>: I had another physician transitioning to practice, this time a neurologist just completing his fellowship.

Just as was the situation for the internal medicine doc, the neurologist's health was sub-par! He had another textbook's worth of medical records. Unfortunately, there wasn't a whole lot we could do. He couldn't covert group coverage to individual coverage without underwriting.

We did shop out his coverage. Unfortunately, none of the traditional carriers would accept him: a flat-out decline. REJECTED!

We even shopped it out to a specialty firm: Peterson. Peterson did make an offer, but... It was an offer that the client could refuse. It covered very little and it *charged more* than a traditional policy. It wouldn't cover nervous disorders; it wouldn't cover kidney issues; it wouldn't cover heart issues; and it wouldn't cover blah blah blah. Basically, the only thing it would cover were musculo-skeletal issues.

The neurologist had group coverage and settled in with it. Unfortunately, sometimes individual coverage isn't worth the premium.

Next, let's review the bells and whistles (riders) that you can add onto your disability income policy.

HERE ARE THE RIDERS THAT I STRONGLY SUGGEST A PHYSICIAN CONSIDER:

- ➢ Own Occupation/Regular Occupation
- ➢ Residual Disability & Recovery Benefit
- ➢ Guaranteed Insurability/Future Purchase Option
- ➢ Guaranteed Renewable and Non-Cancellable
- ➢ Cost of Living Adjustment (COLA)

OWN OCCUPATION/REGULAR OCCUPATION:

For physicians, this is a particularly important rider.

Imagine what would happen if you had to quit your job due to a disability. Maybe you developed arthritis or maybe a horrible pain in your neck forced you to call it quits as a physician, or perhaps limited your hours.

Yet, you want to work in another occupation because the disability income is not enough to maintain your standard of living. (You also wish to do something useful with yourself!) After all, you are used to earning $200,000 to $400,000 per year. How many other jobs could replicate that kind of income?

Without this rider, you could not work another occupation and make a claim on your disability policy. Essentially, you would be solely dependent on the insurance policy.

This rider ensures that if you are no longer able to operate and do procedures as a physician, that you could nevertheless earn income from <u>another occupation</u> (e.g., teaching at a university or working as a financial advisor) in order to supplement your disability income insurance check.

Specifically, a "Specialty Occupation"/"Own Occupation"/"Regular Occupation" rider covers a "job for which you are suitably trained and qualified."

<u>Some of the "Big 6" differentiate between Specialty Occupation and Own Occupation (or Regular Occupation).</u> Truly, specialty occupation does have a more specific definition related to being a physician. However, several brokerage companies I have spoken with have thought that this is merely a marketing gimmick that allows the insurance company to charge higher rates for the rider.

If you feel strongly, if you are more comfortable with and have better peace of mind with one definition versus the other, that's great. Keep in mind that I'm trying to help you keep down your costs.

Either type of rider would work fine; the point is to make sure to include it!

RESIDUAL DISABILITY & RECOVERY BENEFIT:

This is the official name of the rider. I prefer to call it "partial" disability.

As noted in the table comparing group DI versus individual DI, most group policies do not cover partial disability.

This rider ensures that if you can still work as a physician, but if you need to work reduced hours, due to a disability (a minimum of 30% up to 80% of normal hours is typical), for a period longer than your elimination period, then you can receive a check from your disability policy during this period of "partial disability."

If this rider is not included, then any period of partial disability will not be covered.

Ironically, partial disability for less than five years is the most common disability situation. I am sure that this is due to the "stuff happens" in life: the skiing accident that causes you to break a leg or perhaps the years of bending over and overstretching your neck. My dentist recently had to take a full disability and leave from his practice due to the latter issue.

GUARANTEED INSURABILITY/ FUTURE PURCHASE OPTION:

Many physicians start out with a significantly lower income than what they end up earning a few years down the road.

Perhaps they become a partner in the practice in later years, or they are promoted to a leadership position. Also, perhaps they are in a grace period for the first year or two while their clinic tests their mettle to see what they are made of.

Don't Gamble. Whatever the case, the difference in income can be substantial two years or more into practice. While the young physician is paying back student debt and starting to accumulate assets, he or she may not be in a strong enough financial position to gamble on their good health, and back off on their disability insurance- as a matter of fact they may want to even increase the coverage of their insurance.

Can Get Mo'. This rider ensures that you can purchase more insurance for specific increments of coverage at a given point down the road (1 years, 2 years, or 3 years hence, depending upon the company)--without going through underwriting again. Your health will not be considered when you increase your DI coverage as your income increases.

There's a Limit. Note that some companies, such as Principal, will give you the opportunity at a specified time (3 years, in their case) to purchase additional disability income insurance without underwriting. If you choose not to increase your coverage at that time, the future purchase option will cease to exist.

Alternatively, if you think that your income may be capped out and that there is limited likelihood that it will increase, OR if you could maintain your desired standard of living within the limits of your current disability policy, then don't include this rider!

GUARANTEED RENEWABLE AND NON-CANCELLABLE:

These are two separate terms that can be automatically included with your contract or can be added on as separate riders.

Make sure that you understand the difference between the two, as well as whether or not they are automatically included.

If you do not have these riders, the insurance company could cancel your policy (along with other policies in the same age/type category) or could raise your premium at will.

To give you an example, let's review the difference between Principal and The Standard.

I recently ran a quote for a client where Principal and Standard came up as the best two options. Both included provisions for guaranteed renewable and/or non-cancellable.

The Principal contract was slightly more expensive and is BOTH non-cancellable and guaranteed renewable, whereas The Standard contract is slightly cheaper and is ONLY guaranteed renewable.

The difference between the two is the non-cancellable provision, which means Principal cannot change or cancel your policy, except in the case of nonpayment of premiums, nor can the company increase the premiums before you reach age 65, regardless of changes in your income, occupation, or health.

Standard, in contrast, could not cancel the policy (except in the case of unpaid premiums), but the company *could* raise your premiums, if it did so for all people in your age and occupation class.

To quickly sum up: The Standard's contract in this example was cheaper, but premiums could potentially rise in the future, whereas Principal's contract offered a locked-in price, even though the contract was slightly more expensive.

(Note: The Standard allows you to purchase non-cancellable as an additional rider with its Platinum product)

Either one is a fine contract; just be aware of the choices!

CHAPTER 17

THE RESIDENT'S GUIDE TO LIFE INSURANCE
By Dave Denniston, CFA

The next important type of insurance for physicians is life insurance.

Are you engaged or married? Do you have any children?

Imagine what it would be like for your family to lose your income. How would it affect their lifestyle? How long could they maintain their lifestyle without your income?

In the last 15 years of being an entrepreneur, I have seen unexpected death drastically change the lives and fortunes of physicians. It is an incredibly difficult transition time for families. The last thing you want on top of the emotional distress is financial distress.

As a resident or fellow, <u>if you are single</u> and have few to no obligations, then <u>there is little to no need for external life insurance</u> outside of what is provided at work, and you can skim over the rest of the next two chapters. If you are single, educate yourself, but there's no need to put any of these suggestions into action.

However, if you are a resident or fellow and you have a spouse and/or children, I would strongly encourage you to make sure your family is covered!

As is the case with disability insurance, you want to ask yourself, how dependent is my family on my income? Am I a primary breadwinner or part of a high two-income family?

How much in consumer and house debt do I have?

Note that regardless of the way that you hold life insurance, the proceeds are always income-TAX-FREE. The death benefit from a life insurance policy also avoids probate by going directly to your named beneficiaries.

Although I must point out that if the dollar value of your estate is over a certain limit in your state (perhaps as low as $1 million, depending upon the state), or over a certain limit federally (above $5 million, rising each year), then you may owe estate taxes.

There are two basic ways to obtain life insurance: through work (group term) and by funding it yourself (individual coverage).

Just as is true of disability income insurance, there are pros/cons for having life insurance through work and for funding it yourself.

Although this information may seem repetitive, due to the similarities/overlap with disability income insurance, let me review the differences between a typical group policy and an individual life insurance policy.

Typical Group Policy	Individual Life Insurance Policy
Little to no medical underwriting required. Neither age nor health will usually matter at time of application.	Substantial medical and financial Underwriting required. Better the younger and healthier you are.
Cheapest Option (usually a minimal amount of life insurance is free through employers).	Usually more expensive, depending upon the riders you select.
Premiums paid monthly at no extra cost.	Premiums best paid annually. Quarterly or Monthly costs more.
Contract terms, premiums, and benefits can change at any time.	Guaranteed term insurance premiums cannot change (but universal life premiums can).
Policy ends when you end employment.	Policy is carried by you, before and after employment with any employer.

The next most common question is: How much should I have? How can I figure it out?

There are several methods to determine how much life insurance one should have.

The first method is debt pay-off. The objective here is to eliminate all liabilities and to increase cash flow after a loved one's death. This way, your monthly obligations become fairly minimal.

Note that most student loans are NOT included in this calculation, because federally backed loans (i.e., Stafford, Perkins, FFEL) disappear at death. See http://studentaid.ed.gov/repay-loans/forgiveness-cancellation for further details.

I suggest debt pay-off amount as the minimum threshold of life insurance that one should hold.

Also, the debt pay-off method of calculation is ideal for a family in which both spouses work.

Last, I strongly suggest this method for residents and fellows who are looking for affordable insurance, but who do not have much of a budget.

This way, all liabilities are erased in the event of one spouse's death, but you are not over-insured with extra life insurance that you do not need.

The next method is to calculate income replacement. The goal with this method is to replace your AFTER-TAX income for a specified time period. I suggest AFTER-TAX because life insurance proceeds are not subject to income taxes.

Don't Forget. Also, if you are using your last pay-stub for purposes of this calculation, add back any current 401k, 403b, or other retirement plan contributions, and also add back premiums for medical insurance. This will ensure that you are counting the money that you would have been saving for retirement.

How Many Years? Would you like your family to be able to replace 10 years of your income? 20 years? 30 years?

Many folks assume that after 10 years, their spouse will have moved on and landed another job, or will have established another source of income.

Goal Oriented. Other folks want to protect more and to ensure that their family is set for a variety of goals, such as paying for college for the kids.

Some want to be sure that their spouse will never have to work again, and so these folks will focus on replacing their income for a span of 20 years or 30 years.

The Rule. A simple rule of thumb is to multiply your current AFTER-TAX income by 10 to figure out a death benefit.

For example, if your pre-tax income is $180,000 and your after-tax income is $140,000, multiply 10 (assuming 10 years is the target) times $140,000 and this gives you total life insurance needed of $1,400,000.

Are you a resident or fellow in your last year (with a spouse and/or kids) before you enter into full-time practice? I strongly suggest starting the underwriting for this amount in the last few months of your residency/fellowship.

The final method of calculation is the home-sweet-home method. Okay, I am being a little (or a lot?) tongue-in-cheek here, but one could make a case for some folks to add together the debt pay-off method and the income-replacement method. This way, not only are their debts erased, but their income is also replaced for a specified time period. I would suggest that if you go with this method, you use a smaller income replacement time frame than you would do otherwise: think 5 to 10 years rather than 10 years or more, in order to minimize the cost of insurance.

Action Step: How much group term insurance do you currently have from work?

_____

_____

_____

_____

Action Step: Calculate the amount of life insurance that you could need based on the debt pay-off method. How much is [are] your mortgage[s]? Car loan? Credit card debt? Other liabilities? Do NOT include student loans. Do NOT include student loans (unless they are held privately by SoFi/Earnest/etc).

_____

_____

_____

_____

Action Step: Calculate the amount of life insurance that you would need based on the income-replacement method. Make sure to use your current AFTER-TAX income and multiply it times the desired time period. If you are not sure, then use 10 years, 20 years, and 30 years.

_____

_____

_____

_____

_____

_____

_____

_____

Action Step: Calculate the amount of life insurance that you would need based on the home-sweet-home method. Make sure to add together the results from the previous action step and your current AFTER-TAX income, and multiply this amount times the desired time period. Use a smaller time frame of 5 to 10 years in order to minimize the cost of insurance.

_____

_____

_____

_____

CHAPTER 18

WHY CASH VALUE INSURANCE SUCKS (& TERM ROCKS!)

By Dave Denniston, CFA

Now that we've determined the total amount of life insurance that you need to look into, the next step is to understand the difference between term insurance versus cash value insurance.

I will list several examples and then ask you to get some quotes for your particular situation.

Full disclosure: I have a bias that I need to be intellectually honest about. I consider myself a "term-ite" who is addicted to term life insurance because I am cost-averse and like to keep my living expenses low.

Anywho, I will address some cases that I feel are appropriate for cash-value types of life insurance. It's not always a bad decision.

DIFFERENCES BETWEEN TERM AND CASH-VALUE INSURANCE

Guaranteed Term	Whole Life (WL-cash value)	Universal Life (UL-cash value)	Variable Universal Life (VUL-cash value)
Fixed Costs; will not change for designated period of time.	Fixed costs; will not change throughout life of the policy.	Insurance costs can vary based on interest rates and actuarial tables.	Insurance costs can vary based on investments and actuarial tables.
Cheapest premium option.	Expensive premiums.	Fairly expensive premiums; depends on how long policy exists.	Fairly expensive premiums; depends on how long policy exists.
Does not have cash value.	Does accumulate cash value.	Does accumulate cash value.	Does accumulate cash value.
No surrender charges, due to no cash value.	Long surrender charges: Must hold policy 10+ years in order to cash out.	Long surrender Charges: Must hold policy 10+ years in order to cash out.	Long surrender charges: Must hold policy 10+ years in order to cash out.

In comparison to rating systems for disability income insurance where they focus on your occupation, a life insurance company's premium that is offered to you can vary widely based on a number of different criteria related to your health and history.

DI Is Different. For example, let's say that you have a history of a mental health disorder. A disability income policy would likely EXCLUDE any claims for the disorder, but the premium is still the same.

So, you might get a condition excluded- but they give you the same premium as someone in similar age/occupation.

LI Changes The Game. In comparison, life insurance rates individuals based on their health. The poorer your health, the more expensive the life insurance will be. This is regardless of the type of life insurance.

The best possible health rating is super preferred, followed by preferred plus, then preferred, standard plus, standard, and then tabled ratings.

Can't Get Tabled. A tabled rating is insurance for those with chronic conditions or those whom the insurance company would deem to be high risk.

Short & Sweet. Also, when it comes to term insurance, the longer the term, the more expensive it is. The shorter the term, the cheaper it is. You could be extremely cheap and buy 10-year term, or you could pay a little bit more and get 20-year term, or you could pay the most for 30-year term.

Keep in mind that the premiums will increase after the designated period. Work towards having insurance until the time you calculate that you should be close to self-insured or, alternatively, near retirement.

In addition to the differences mentioned between term and cash-value insurance, let's walk through some examples so that you can further understand how this could work in real life.

I ran some quotes for a 32-year-old for $1,500,000 in life insurance death benefits, and entered the results in the tables below.

10-Year Term, Pref Health	20-Year Term, Pref Health	30-Year Term, Pref Health	Universal Life, Pref Health
$495/ year	$870/ year	$1,380/ year	$3,340/ year

20-Year Term, Pref + Health	20-Year Term, Pref Health	20-Year Term, Standard + Health	20-Year Term, Standard Health
$615/ year	$870/ year	$1,050/ year	$1,350/ year

See how the longer the TERM insurance goes, the more expensive it is?

A 30 year term policy is more expensive than a 20 year term policy which is more expensive than a 10 year term policy.

In drastic comparison, cash-value insurance is the most expensive because it is supposed to cover the insured person until you are at least 90 years old (or maybe even beyond!).

Thus, the 10-year term policy is only about 85% cheaper than is the universal life policy!

Also, see how the stronger your health, the cheaper the premium?

Using a 20-year term policy as a reference, we can see that premiums for a preferred plus health rating are 50% cheaper than are those for a standard health rating.

Although I am an admitted "term-ite," there is a time and a place for cash-value life insurance.

Scenario 1: Let's say that you are maxing out your employer-sponsored retirement plan, such as a 401k or a 403b plan, and on top of that you are maxing out your Roth IRA (or back-door Roth IRA). You may have $1,000,000 plus there.

In addition to that, you have saved non-retirement funds of $500,000 or more, which is enough to cover years of living expenses. Also, let's assume that you have paid off all of your student loans, high-interest-rate, non-deductible debt, & your tax-deductible interest via your mortgage.

This could be a time to consider cash-value life insurance for a portion of your savings. However, keep in mind that the older you get, the more expensive life insurance becomes.

Adding more to your non-retirement account or purchasing a tax-deferred annuity may be a better fit for you in this instance, if you would like tax deferral and/or an insurance-company-guaranteed income stream down the road.

Scenario 2: A very strong case can be made for using cash-value life insurance as a tool for estate tax planning. Using a cash-value life insurance policy owned by an irrevocable life insurance trust (ILIT), you can separate out the death benefit of the policy from the rest of the estate.

Why would you want to do this?

Estate taxes! You could potentially save hundreds of thousands of dollars in estate taxes by using an ILIT with a cash-value life insurance policy.

For example, the state of Minnesota currently has a newly $2,000,000 estate tax exemption. This means that Minnesota will require the estate to file an estate tax return if any assets in the name of the deceased exceed $2,000,000.

This asset amount INCLUDES life insurance proceeds, 401k accounts, investment accounts, bank accounts, land, etc.

So, if a deceased person has a $1,000,000 life insurance policy and a $1,500,000 401k, then that person's estate totals $2,500,000, and the estate would be taxed on the excess above $2,000,000: in this case, $500,000. In the state of Minnesota, taxes start at 10% and go up from there. This hypothetical estate of $2,500,000 would be hit with an estate tax bill of at least $50,000 (10% of the "excess" $500,000).

Note that each state has different rules and regulations.

For example, community property states such as Texas and Washington divide up the spousal assets equally when the first spouse of a legally-married couple dies. Both spouses face the same exact estate issues.

Minnesota, on the other hand, looks at the assets in each spouse's name and keeps these assets separate. One surviving spouse could potentially face an estate tax issue while the other does not.

On top of that, there are also federal estate taxes. The exemption in 2016 is close to $5,500,000. As long as the value of your estate is below that, you don't need to worry about federal estate taxes. However, any estate valued above that amount is subject to a very steep tax of 40%. Keep in mind that the value of the estate does include any life insurance death benefits (unless the insurance is owned by an ILIT).

I know, it's weird! But that's why you need to consult with a financial advisor and/or an attorney if you think you could be affected by these considerations.

Action Step: Get quotes from several companies based on your ideal amount of life insurance, to fill in the table below.

10-Year Term, Preferred Health	20-Year Term, Preferred Health	30-Year Term, Preferred Health	Universal Life, Preferred Health
10-Year Term, Standard Health	**20-Year Term, Standard Health**	**30-Year Term, Standard Health**	**Universal Life, Standard Health**

Action Step: Re-visit Scenarios 1 and 2 to see if it may make sense to look at a cash-value life insurance policy. For example, calculate the current amount of your estate. Is it worth looking into a cash-value irrevocable life insurance policy?

_____

_____

_____

CHAPTER 19

WHAT IS INVESTING?
(HINT: NOT YOUR CAR)
By Dave Denniston, CFA

In the rest of the investment section that is coming, we are really going to get knee deep into examples and numbers and strategies.

If you are an intermediate or advanced investor, skip ahead to the next chapter.

If you are like most residents and you are getting started on your journey, you may not have had much exposure to investments. This chapter is for you!

You know that investing is a good thing, but you're not familiar with the terms and it feels a bit over your head.

You may even feel a bit embarrassed when your financially knowledgeable colleagues discuss 401ks, IRAs, Roths, ETFs, and mutual funds and you have a concept of what they are talking about, but you aren't really sure what that all means.

Let's clear that up right here and right now and get you up to speed.

Most investments really come down to two things- equity or debt.

WHAT IS AN EQUITY INVESTMENT?

Equity is simply a fancy term for ownership. You may own a large piece or a small piece of an asset.

That asset could be any of the following and much more:

- A piece of real estate
- A stock
- Your future private practice
- Other privately held entities

When you have equity, you have an ownership stake in an investment.

When we discuss stocks, we are talking about equity. Your ownership piece of that asset may be really, really small in the case of companies that are traded publicly- like Coke, Microsoft, or Google.

As an owner of an asset, the asset can skyrocket up in value or it can go bankrupt. Many ownership assets tend to be very volatile- they move up and down in value quite frequently and with significance.

Your ownership investment may pay a cash dividend…. or it may not. The dividend is a form of paying shareholders a part of the profits of the company. (Although, some companies pay dividends even though they don't have profits.) Basically, there's no obligation of the company to continue to pay dividends.

When we talk about stocks and your ownership of them, we often refer to the individual unit as a 'share'.

How do you measure return of your equity investments?

You focus on two things- growth (loss) of the principal and the dividends you receive.

This is the total return.

If you lost 3% of principal, but had 6% in dividends, you are still up 3% overall.

Always keep total return in mind as you examine dividend paying entities in comparison to non dividend paying entities.

WHAT IS A DEBT INVESTMENT?

In comparison, debt investments are the opposite of what you have been doing with student debt payments.

When you took on student debt, you owed a company a mountain of money and you will pay them huge sums in principal & interest.

When you invest in debt, you are flipping that concept around. You are NOT borrowing money. You are lending money.

When you invest in debt, someone else owes you money- principal and interest.

This debt could be on real estate, equipment, or just on the general financial ability of a company.

Real estate and equipment can be considered collateral. This means that you could go and snatch those assets if the person you lent money to doesn't pay the principal and interest like they are supposed to.

Many loans have no collateral, but if the company goes bankrupt, you can be one of the first people in line to collect money when the company's assets get liquidated.

We see debt investments in the form of:

- Bonds
- Peer to Peer Lending (Lending Club/Prosper)
- Real estate loans
- Checking/Savings
- CDs (Certificates of Deposit)
- Money Market

Yes, you read that right. Checking, Savings, CDs and money market are a form of debt investments.

The bank is paying you interest because you are loaning them money.

If you look on financial statements for a bank, their assets are the loans that they give to consumers (car loans, house loans, business loans, etc).

In contrast, their liabilities are the deposits at the bank. They are borrowing money from you in order to create assets where they let other people borrow from them.

WHY DO WE INVEST?

It's a simple question, but it's worth some pause.

Why do we invest?

We want to make money. Absolutely.

But there are all kinds of ways of making money.

I assert that we invest because we want to reach our goals faster. We want our assets to grow faster. We are chasing the comfort and the life that we dream of. We invest to provide for our family in the future so we don't have to work forever.

Most of all, we don't want to have to do the labor of working and instead 'passively' invest allowing our money to work for us rather than us working for money.

THE RULE OF 72

How can we measure our pace and our returns?

There's a simple investment rule called the Rule of 72. It is a basic way of figuring out how long it will take to double your money.

Later on in the text we will discuss risk and reward. The Rule of 72 serves as a great foundation for understanding the consequences of risk & reward.

What you do is you take 72 and divide it by a rate of return (or desired rate of return) on your investments.

This will tell you how long it will take to double your money.

Let's take a look at a couple of examples.

EXAMPLE# 1: RATE OF RETURN OF 12%

If you have a rate of return of 12% over a year, you take 72 and divide it by 12. This gives you 6.

Assuming you average a 12% rate of return over the next 6 years, you will have doubled your money.

Your $100,000 would have become $200,000 without putting in any additional principal.

EXAMPLE# 2: RATE OF RETURN OF 6%

If you have a rate of return of 6% over a year, you take 72 and divide it by 6. This gives you 12.

Assuming you average a 6% rate of return over the next 12 years, you will have doubled your money.

<u>This simple example shows us the power of investing. By taking risk with your money and earning higher returns than by stuffing the cash in your mattress, you can hit your goals much sooner and even better you can quantify how you can get there.</u>

Ultimately, it all comes back to your goals. You can retire at 55 instead of 60. You can do more to help your kids through college. You can take the extra trip to Europe or splurge on a better car when you hit your goals.

You can give more to your church or your faith organization.

<u>This is why we invest.</u>

A QUICK PRIMER ON DIFFERENT KINDS OF ACCOUNTS

There are all kinds of ways you can invest. We're going to try to make it as easy as possible.

Pre-Tax Now, Post-Tax Later. You get a tax deduction for putting money in, but when you pull the money out it is THEN taxed. This is not so important in residency. However, it can be very important when you are in practice.

Once you put the money in, it is meant to be there until you are about 60 years old. You are not meant to touch it. Of course, there are ways you can- but I generally recommend seeing this money as untouchable.

In the years between when you invest the money and when it comes out, you don't have to pay any taxes on your gains or dividends or interest.

This money is solely in your name. Your spouse or other designated person can be the beneficiary.

Pre-tax now and post-tax later accounts are IRAs, 401ks, 403bs, 457 DCs, and other similar ones.

Post-Tax Now, Tax Free Later. You don't get a tax deduction for putting the money, but when you pull the money out- it is tax-free. This IS important in residency when you are in a relatively low bracket and can also be important when you are in practice. See the Freedom Formula for Physicians (www.doctorfreedombook.com) for instructions on the back-door Roth IRA if you are already in practice.

Like the pre-tax now, post-tax later world, these accounts are generally not supposed to be accessed until 60 years old. However, there are more friendly ways of getting your hands on this money- particularly the principal that you invested with. Although, I generally recommend seeing this money as untouchable as well.

Just like our first example, in the years between when you invest the money and when it comes out, you don't have to pay any taxes on your gains or dividends or interest.

Like our pre-tax now, post-tax later category, this money is solely in your name. Your spouse or other designated person can be the beneficiary.

These kinds of accounts are Roth IRAs, Roth 401k, Roth 403b, etc, etc.

Post-Tax Now, Pay Taxes As You Go. This category is what we call non-qualified money. There is no tax deduction for putting the money in. There is no huge tax advantage for taking the money out.

It's simply like a bank account. You can pull out the money at any time.

You can get taxed on your annual interest, dividends, and capital gains.

This money can be in your name, with your spouse, just your spouse, with your kids, with your parents, in your corporation's name, or however else you may want to title the money. It could even be in a trust of some sort.

I generally suggest this kind of account to add to your cash cushion money. It is liquid and easy to get to, but it gets taxed as you go.

DOLLAR COST AVERAGING

Our last concept to comprehend is this idea called 'dollar cost averaging'. Essentially, this is a fancy financial term that ensures you are investing every single month.

Imagine two different doctors. Dr. Smith and Dr. Jones.

The Tale of Dr. Smith. Dr. Smith is scared of investments like stocks and bonds and doesn't want to lose any money. In addition, he likes to have a big cash cushion at the bank.

He is a saver and loves seeing his balances get bigger and bigger. Dr. Smith builds up his cash cushion month after month. On average, he is saving $1,000 a month. Over the course of a year, he has saved up $12,000.

He finally decides to invest some money and does it in one lump sum. He wires money to his brokerage account and invests. He continues to save $1,000 a month in the bank while the investment grows.

Over the course of the year, the combined savings grows to $25,000-$24,000 of principal and $1,000 in growth (8% on the $12,000).

The Tale of Dr. Jones. Let's compare this to Dr. Jones. Dr. Jones isn't scared of market risk and wants to 'put his money to work' as soon as he can.

Rather than bank $1,000 a month, he invests it using dollar cost averaging. He invests every single month at the same exact time like clockwork.

Some months he is down and some months he is up. However, over time, he is up about 8% a year.

Dr. Jones' journey may have looked like this in year one:

Month	Shares Bought	Total Shares	Ending Value
1	100 @ $10/share	100	$1,000
2	105.26 @ $9.5/share	205.26	$1,949.97
3	108.11 @ $9.25/share	313.37	$2,898.67
4	111.11@ $9/share	424.48	$3,820.32
5	108.11 @ $9.25/share	532.59	$4,926.46
6	105.26 @ $9.5/share	637.85	$6,059.58
7	102. 56 @ $9.75/share	740.41	$7,219.00
8	100 @ $10/share	840.41	$8,404.10
9	97.56@ $10.25/share	937.97	$9,614.20
10	95.24 @ $10.50/share	1,033.21	$10,848.69
11	93.02 @ $10.75/share	1,126.23	$12,107.00
12	92.59 @ $10.80/share	1,218.82	$13,163.26

At the end of year one, Dr. Jones who dollar cost averaged is already ahead of Dr. Smith who held money at the bank and waited for it to pile up.

Dr. Jones had $13,163.26 while Dr. Smith had only $12,000.

You can imagine how over time- this small difference compounds and compounds some more. Dr. Smith is chronically behind Dr. Jones until he happens to time the very market bottom and invest in the right 3 or 4 months.

The Lesson To Learn. My friends, the lesson here is that it doesn't matter whether the markets are up or down. You invest steadily. You buy more shares when assets have been beaten up. You buy less shares when assets have appreciated.

The concept is that you don't have to time the markets. Dollar cost averaging will allow you to buy rain or shine so that you get invested.

Residents and fellows and newly practicing physicians would be very wise to get started with some dollar cost averaging.

It doesn't have to be much. $50 or $100 a month will suffice to get you started.

It could be in your 403b or 401k or a Roth IRA or a non-qualified account just in your name. The point is get started as soon as possible!

CHAPTER 20

INVESTING 101
By Dave Denniston, CFA

THE BASICS

I was raised in Southern California. In the nearly 18 years I lived there, both football teams moved away, and we were left solely with professional basketball and professional baseball.

The allure of the Showtime purple and gold caught my imagination as a kid. Magic, tossing unbelievable behind-the-back passes. Run and gun, fast-break basketball.

Yup, I'm a Lakers fan.

Fast-forward 10 years, and I'm in college at Seattle Pacific University. During the Shaq and Kobe years, I was so caught up in ballin' that I watched the NBA like the superfan I was.

Hey, we won three championships in a row—can you blame me?

I watched LOTS and LOTS and LOTS of NBA ball. I watched TNT pretty much every night. Saw every highlight reel.

I delighted in the drama.

One quote stuck with me—I laugh, giggle, and snicker every time I think about it.

"We're talking about practice man, practice. We're talking about practice. We're talkin' about practice, man."

Best. Quote. Ever.

Allen Iverson was mocking his coach's insistence on practicing when he had tore it up in games. Why should he need to practice?

Well, you know what? "We're talking about investment basics, man, investment basics. We're talking about basics."

This chapter is going to focus on fundamentals. Whether you are a beginning investor or a more advanced physician investor, there are strategies here for you. We'll start with the basics, and then move our way toward more advanced strategies.

In my other book, *The Freedom Formula for Physicians* (www.doctorfreedombook.com), I outline more advanced strategies for the adventurous who want to know one of the ways I personally use to time the market. They can be extremely inappropriate for any young physician, but you should be aware of the pros and cons.

STEP 1: KNOW YOUR RISK TOLERANCE

The first step you must take is knowing your risk tolerance. You've got to grasp how much risk you are willing to take.

There are several basic determinants that drive risk tolerance:

✓ Time horizon
✓ Amount of funds needed
✓ Length of time funds will be drawn
✓ Amount of liquid & emergency resources
✓ Planned big ticket purchases (cars, houses, etc.)
✓ Loss aversion & behavioral finance
✓ Goals

College and Retirement. For example, let's say that a physician is planning on saving simultaneously for their kids' college as well as for retirement.

Retirement is 20 years away and will probably last for another 20 years.

Let's take a look at another scenario. Let's imagine that there are funds saved up for college. College for the kids is two years away, and that money will last for about four years. It's a much shorter time horizon.

Could you see how risk tolerances could be different for these situations?

If the markets have a really cruddy year, the retirement account has PLENTY of time to recover, but the college funds in this scenario DON'T.

The risk tolerance could and likely SHOULD be different.

THE 50% DRAWDOWN RULE

Here's a simple example I like to draw for folks.

Let's say that you start with $100,000.

You invest it really aggressively, and have a really, really bad year of -50%.

How much money do you have now?

. . ...

. . ...

Yes, $50,000.

<u>Now, here's the kicker- what percentage do you need to get back to</u> <u>$100,000 from $50,000?</u>

If you lose 50% of $100,000, you now have $50,000.

You say, "Duh, Dave! I knew that!"

Okay, keeping reading. Consider this . . .

<u>You will need a 100% rate of return to get back to $100,000—</u> <u>$50,000 over $50,000!!</u>

You lose 50%, but need 100% to get back to where you started.

Let's take another example. Let's say your $100,000 loses 20% instead.

Your $100,000 became $80,000. You need $20,000 over $80,000 to get back to where you started.

That's 25% rate of return. Imagine how much easier it is to regain what you've lost than the 100% in the previous scenario.

The worst time to change your risk tolerance to be more conservative and sell is when the markets have dropped like a rock. I don't mean a small 5% or 10% blip. As we've just discussed, that's not a big deal and is much easier to overcome.

THE MISTAKE THAT IS INCREDIBLY HARD TO OVERCOME

Consider that at the depths of the financial crisis stocks were down about -50% in 2008 through the beginning of 2009.

A scared investor suddenly changes their risk tolerance from an aggressive mix to a conservative mix.

Think about this for a second . . .

They've cut their stock exposure in half, and now have to get 100% rate of return with a fraction of the ability to get those returns.

It could literally take twice as long.

AN EVEN WORST MISTAKE...

Let's take a look at another example where NOT ONLY have you LOST money, but NOW you are withdrawing money.

Your $100,000 became $50,000.

Reimbursements have been declining. The profitable procedures have been less and less profitable. Unfortunately, you don't get the productivity bonus you were counting on.

Then, you need some cash flow, and take out $10,000 from the remaining $50,000 in your account. Your $100,000 quickly became $40,000.

<u>Now you need a 150% rate of return (60,000 divided by 40,000) in order to get back to where you started.</u>

Folks, this is how I have seen people literally go broke.

Withdrawing from your portfolio in a down market can be VERY, VERY painful—perhaps irreparably so. I can't emphasize this enough—be very careful of your risk tolerance, and make sure you are committed to staying that course until the time comes for gradually shifting as your time horizon shifts.

THE WORST INVESTMENT MISTAKE I MADE

I came into this business during the tech crash of 2000 to 2002. I was trained as an intern, by this vibrant financial advisor, Vivienne Strickler.

Vivienne was amazing and a great mentor. However, she preferred mutual funds, and when I looked at the internal costs, I wasn't all that impressed. Why couldn't I build my own stock portfolio, and try to beat the market and cut out that expense?

Coming out of the tech crash, I did some pretty smart things. I studied financial statements. I bought beaten-down stocks. I went with mega-sized, dividend-paying stocks that could maintain the dividends for a long, long time. I bought big, large cap names like McDonald's, Starbucks, Duke Energy, and Nvidia.

THE WORST INVESTMENT MISTAKE, CONTINUED

All of them went up by at least 30%. Nvidia even doubled for me. I was making a bunch of money, and I thought I was pretty hot stuff.

The idea was to buy individual securities that'd had the snot beaten out of them: names that were fairly well known, and had some form of security behind them, and strong financials.

I kept at this same strategy. Then, one day, a delicious, delectable, sweet treat caught my eye. Glazed with sugary, sweet goodness. It also captured my heart through my stomach. Yes, the Krispy Kreme.

It was a hot, hot stock. Soaring away like the newest Starbucks. A friend told me, *Eat the donuts and not the stock.* I listened and held off.

Then, it crashed. And crashed yet again. It started to rise again, and I snatched up some shares. I also bought some for my mom. I was so sure this was a sure thing. Unfortunately, I was wrong. Dead wrong. It continued to sink and sink some more.

I held on for a whole year, and it became clear that it wasn't coming back. I sold it at an 80% loss. <u>In the meantime, the market rose double digits</u>. It was my worst mistake. Ever.

The lesson I learned: I cannot beat the market with individual security selection—it could vary way too much from the market.

DECIDING ON A RISK TOLERANCE

There are five basic risk tolerances most folks use. On a scale of 1 to 5:

➢ Conservative
➢ Moderately Conservative
➢ Moderate
➢ Moderately Aggressive
➢ Aggressive.

Each of these has some volatility, right?

You'll go up and down on a daily basis. It's just a matter of how high, how low, and how quickly.

Here's a typical asset allocation mix for each of these risk tolerances:

Risk Tolerance	Stocks or Riskier Asset %	Bonds or Less Volatile Asset %
Conservative	40%	60%
Moderately Conservative	50%	50%
Moderate	60%	40%
Moderately Aggressive	70%	30%
Aggressive	80%	20%

Here's my challenge to you:

• Do you know your risk tolerance?
• Are your investments in line with your risk tolerance?
• Do you know when you will shift your risk tolerance?

To help you better understand what volatility means in the real world, I did some research into the high points and low points for standard portfolios with BOTH domestic and international stocks over the last 10 years. (See chart.)

Portfolio	10 Year Average	Worst One Year	Best One Year	Best 3 Years	Worst 3 years
Conservative	6.08%	-23.35%	35.62%	17.5%	-5%
Mod. Conservative	6.61%	-26.33%	40.40%	19.87%	-6.77%
Moderate	6.53%	-34.12%	51.78%	23.69%	-9.17%
Mod. Aggressive	6.33%	-34.88%	48%	22.67%	-9.58%
Aggressive	7.34%	-39.3%	56.31%	26.1%	-12.19%

Source: Morningstar Advisor Workstation, Prior performance does not indicate future performance, data through 8/31/15

In this table, a number of interesting characteristics are revealed.

Conservative Results. First, the conservative portfolio actually did pretty well—it averaged about 6% over the last decade. However, the equity exposure can still make for some bumps and bruises.

The worst year was -23.35% and its best year was up +35.62%—a dispersion of nearly 60%.

Aggressive Results. In stark contrast is the aggressive portfolio, which you have the aggressive portfolio, which added nearly 1.5% additional return every year, with significantly more volatility.

Its worst year was -39.3%, and best year of +56.31%—a dispersion of nearly 100%.

Folks, this means the road will be bumpy—no matter your risk tolerance. It's just a matter of how wild a ride you are willing to take.

Note: International investments drastically affected the results. We discuss more about the pros and cons of international investments on Chapter 23.

Further Note: This is all backward-looking data. Bonds and stocks will perform differently in the next decade. For this reason, we have assumed a portfolio of a variety of credit quality for bonds.

FINAL THOUGHTS

The key to financial success in the markets, time and time again, is diversification. The best way to determine your mix of diversity is by knowing your risk tolerance. That, my friends, is Investments 101.

RESOURCES MENTIONED

The Freedom Formula for Physicians:
www.doctorfreedombook.com

The Truth About Annuities:
www.doctorfreedompodcast.com/annuities

Morningstar (for independent analysis):
www.morningstar.com

CHAPTER 21

WHY BONDS?

By Dave Denniston, CFA

In Chapter 24, we introduced the concept of equity vs. debt and stocks vs. bonds. Further, in Chapter 25, we showed some basic allocation strategies.

You may be sitting there wondering, "Why the heck should I have bonds in my portfolio?" You may have heard a lot of concern about bonds lately as interest rates have slowly started to rise.

I believe that still bonds are an important part of most people's portfolios.

Let me explain.

They have been the safety valve. Bonds have provided a cushion against the volatility of the stock market.

This is because bondholders are the first investors in line when a company goes belly up. <u>Meanwhile, stockholders lose everything in bankruptcy.</u>

Bond interest is one of the last things to get cut (although they can get called), but stock dividends can get slashed on a whim of the board of directors.

Basically, bonds usually have more safety, and usually have a higher yield than stocks.

Institutional money flocks to high-quality bonds during a downturn.

This was no surprise in the last few decades.

For example, during the last financial crisis through the end of 2009, bonds had been a winner over a WHOLE DECADE, whereas stocks had been a loser.

Yet, the landscape is shifting as we look forward to interest rates rising over the NEXT DECADE.

In May 2013, Fed chairman Ben Bernanke made an announcement that the Fed was merely STARTING to think about removing bond-buying programs and raising short-term rates, bond yields skyrocketed in a very short time period.

This got many folks shaking in their boots about what to do with bonds. I admit it, I was quivering some, too!

First, let's explore the most important aspects to be aware of with bonds.

- ➢ **Credit Quality**
- ➢ **Duration**
- ➢ **Issuer**
- ➢ **Currency Risk**

In my opinion, credit quality and duration are the two MOST IMPORTANT qualities to examine when it comes to any investment vehicle, exchange traded fund (ETF), or mutual fund.

Let's break down credit quality.

Description	Classification	Interest Rate
AAA, AA, A, and BBB	Investment Grade	Very Low to Fairly Low
BB, B, CCC, CC, C	Hi Yield Or "Junk" Bonds	Medium to Fairly High
Not Rated	Crazy Junk Bonds!	Crazy High!

<u>Basically it boils down to this: Do the rating agencies think the company (or country or entity) is going to go broke?</u>

A company gets a super high, pristine rating when they have little debt and lots of cash. *<u>Essentially, there's no chance they will go broke.</u>*

The closer companies are to being broke and bankrupt, the higher the interest rates demanded by the markets.

DURATION

In comparison, duration measures the volatility of bonds approximately via changes in price. It is linked to the total maturity and features of the bonds.

Let's break this down in an upward sloping yield curve, the kind we have had the last decade.

Duration Measurement	Classification	Interest Rate
0 to 5	Low Duration	Very Low to Fairly Low
5 to 10	Intermediate	Medium
10+	Long-Term	High

A common investment that many investors flock to are US Treasury bonds (aka Treasuries). These are bonds backed by the faith and paying ability by the US Government.

Despite many concerns about how good or bad that may be in the future, today- these are the most common bonds that are on the market today. Millions and billions of dollars of these bonds change hands daily.

Let's explore how Treasuries could react if interest rates moved up.

Here's an example:

Name of Option	Duration	Yield
1 to 3 Year Treasuries	1.87	0.60%
7 to 10 Year Treasuries	7.6	1.79%
20 Years+ Treasuries	17.66	2.2%

Source: Morningstar.com

As of the time of this writing (August 2016), bond yields surprisingly lessened since the end of 2013, and 10-Year U.S. Treasuries hovered around 1.6% to 2.2%.

How does this compare to a few years ago?

Consider that back in June of 2008, 10-year treasury bonds were yielding 4.3%. *That's a 2% difference from where they are today.*

How would this play out?

What would happen if 10-year U.S. Treasury bonds' yields increased by 2%, back to where they were a few years ago?

Through the curriculum provided by the CFA institute, if we take the duration x the change in yield, we get an approximate change in price.

For example, if you take a 10-year-maturity treasury bond, duration is usually around 8.

If yields move up 2%, we multiply duration of 8 times -2%.

That means a potential 16% negative change in price!

Alternatively, if 2-year bonds with a 1.5 duration move up 2% in yield, the price only changes -3%.

Yet again, if 30-year bonds with a 25 duration move up 2% in yield, the price change is -50%!!!

Thus, low duration is one way to protect yourself in a rising rate environment. You give up some yield, but keep the protection against rising rates.

Yet, the question is: How quickly will rates rise?

2% change in a year would cause a bleeding gash in a bond portfolio! However, if it was spread out over 5 years—or even better, 10 years—that's not so bad for intermediate bonds.

This is a very present time concern. What if yields continue to move up?

FINAL THOUGHTS ON BONDS

I expect treasury yields to be shifting a bit.

The question for investors and financial professionals to grapple is this: How are we going to handle this?

The answer, in my opinion, is to look to diversify further.

If yields go back to what they were nearly 10 years ago, and you are invested in maturity lengths of 5 to 10 to 30 years, you could get slammed as interest rates move up.

Instead, in my opinion there are three specific areas in bonds that are worth considering.

-   **High Quality, Higher Yield.** Invest in good quality—A and BBB—corporate bonds that aren't quite the high-yield, lower-quality bonds, but aren't the highest quality, either.

-   **Keep It Low.** Also, invest in low-duration bonds to make sure you don't get creamed by the volatility in long-term bonds.

-   **Low Quality, Highest Yield.** In addition, consider adding into your portfolio lower-credit quality, higher-yielding corporate bonds, ones that aren't as safe, **but give a higher yield.**

However, be careful and don't put too much there. This is because higher yielding bonds are more sensitive to changes in the economy than investment-grade bonds. This is because they are considered a poorer credit risk, meaning that they are more likely to become bankrupt and can be a more volatile investment.

A few years down the line, also consider diversifying against the dollar. I expect that there will be a time when the dollar may no longer be a reserve currency.

This means you may want to consider international bonds as well as domestic bonds at some point in the future.

I don't know when this may happen—it may be 3 years, 4 years, 5 years, or 10 years away, I don't know.

However, I don't think this is going to happen in the next year, and this isn't a present time concern.

Consider this question . . .

What will happen if the dollar is no longer the reserve currency for the world?

Japan and China have already started rattling their sabers.

When this momentous event occurs, we should expect treasury yields to really shoot up.

It may be 4 or 5 or 10 years down the line, but at some point, diminishing U.S. power will come home to roost.

For example, what will happen when the Chinese take the punch bowl away, and reduce their bond purchases, in order to create more internal consumption?

We could consider adding non-dollar denominated exposure, particularly in countries that are on the rise economically, as well as those that have strong credit ratings.

There are a number of different ways you can protect yourself in a rising interest rate environment. You don't want to wait until it has already happened.

Note that my personal opinions and outlook may not come to pass, and that past performance is no guarantee of future results.

RESOURCES MENTIONED

Morningstar (for independent analysis):
www.morningstar.com

The Aggregate Bond Index:
https://index.barcap.com/

Definition of Duration:
http://www.investopedia.com/exam-guide/cfa-level-1/fixed-income-investments/duration.asp

CHAPTER 22

THE WIZARDS OF WALL STREET (OR NOT!)
By Dave Denniston, CFA

As an entrepreneur & risk taker, what's fascinating to observe is how unrealistic most folks are. They think that brokers are the wizards of Wall Street (or that they can't get it wrong if they manage the investments themselves). It's as if they think that we have a magic wand that can take risk away.

Most folks want:

1. All the upside of the markets,
2. None of the downside of the markets,
3. Someone to tell them it will be okay when these two things don't happen.

Most advisors will have their clients fill out a risk tolerance questionnaire. This may be 5 to 20 questions and gets to two simple questions.

1. How much are you willing to lose, and how will you act when that happens?
2. How much do you want to gain?

<u>What's ironic is that behavioral finance tells us that this shifts.</u>

People want to follow the crowd. They don't want to lose out. They want to take MORE risk at the WRONG time.

Yet, when there is loss, the PAIN of loss causes people to go nuts. They want to take LESS RISK when everything has gone to hell in a handbasket. This is also the WRONG time.

What we in the industry don't talk about enough is that risk tolerance can be SITUATIONAL.

We have people fill out questionnaires, but don't truly understand how they may react in any given situation. Will they freak out when the market goes down?

Thus, I suggest as a benchmark—you may want to take your risk tolerance DOWN by a notch from wherever you score—particularly if the markets have had a BIG run up.

If you score aggressively, think really hard about how you would feel in a year that you are down -40%.

Are you committed to holding on, no matter what the markets do?

Can you ignore the monthly statements that comes in?

Or would it be so painful that you can't take it anymore?

<u>Think about it . . . how would you react?</u>

ETF VS. MUTUAL FUNDS

I have to confess something to you.

I am a nerd. I love sci-fi movies and books and history!

I'm a student of my industry—and there are some interesting facts we can all learn, and keep in mind.

Back before the Great Depression, the financial markets were about single issuers—single stocks and single bonds.

You HAD to go through a broker. The broker had to place trades through their company's floor trades. The broker and the trader were both employed by the same major Wall Street firm.

In those days, the broker would buy and sell inventory of the stocks and bonds the brokerage firm held. The broker might buy General Electric or Coca-Cola or United Pacific. However, they were often constrained by the inventory, and investors wondered if they had their best interests at heart.

Essentially, you had to create your own stock portfolio to develop some degree of diversity.

The problem was that it was often hard for individual investors to invest in the markets. You pretty much had to be wealthy to avoid extremely high costs of doing business.

The individual investor was consistently getting screwed! (I know, you think this is still the case today. Perhaps some things don't change!)

The Advent of Mutual Funds. Small and large investment firms came up with a different concept—the mutual fund. The basic idea of a mutual fund is that a whole bunch of investors pool their money mutually to invest across a diversified portfolio.

<u>Basically, you can invest a small amount of money to hold many different positions.</u> This simply wasn't possible before.

Investopedia noted that, *"By 1929, there were 19 open-ended mutual funds competing with nearly 700 closed-end funds.* (Note: Open ended mutual funds don't trade on a stock market exchange and constantly buy and redeem shares. Whereas, close-ended funds do trade on a stock market exchange and raise money periodically. They don't handle redemption since they are traded on an exchange.) *With the stock market crash of 1929, the dynamic began to change as highly-leveraged closed-end funds were wiped out and small open-end funds managed to survive."*

This concept was slowly adopted by the general public.

Investopedia continued, *"At the beginning of the 1950s, the number of open-end funds topped 100. In 1954, the financial markets overcame their 1929 peak, and the mutual fund industry began to grow in earnest, adding some 50 new funds over the course of the decade."*

The huge majority of mutual funds at this time were "loaded" mutual funds, meaning that you pay an upfront charge to buy it.

Sky-High Commissions. Back then, these were extremely high! 8%, 9%, and 10% were all fairly standard commissions.

You really HAD to hold it for a super-long time, or you would likely sell at a horrible, atrocious loss due to the commissions you would pay for BOTH buying and selling.

The concept of paying a commission upfront- 3% and up- for a mutual fund is called a 'loaded' fund.

In the 1970s, the first index no-load fund came into being. This meant that there was NO commission paid. It was like pocketing 6%,

7%, 8% up front. You got to keep ALL your money and solely had market risk.

The revolution began! The rise of individuals being able to manage their portfolios became possible. They no longer needed a broker.

This concept was heralded by Vanguard and Jack Bogle with the advent of The Vanguard 500 Index fund in 1976.

Since that time, mutual funds have evolved, and the platforms that they are sold on have changed.

Hidden Fees. However, the way that asset managers run the fund hasn't changed all that much. This is where folks need to understand that most mutual fund fees are HIDDEN.

Here's the deal, guys.

You don't write out a check to the mutual fund manager every month for managing the money.

You don't write out a check to pay for the operating expenses.

You don't even write out a check to the accountant of the mutual fund.

Instead, what they do is take money out of the mutual pool. They take it out of your investment, as well as out of your fellow investors' investment.

Here are the fees you need to be aware of with mutual funds:

Name of Expense	Brief Description
Management Fee	How the managers get paid
Administrative Fee	The cost of tax preparation / documentation / annual reports. Sometimes this is rolled into the management fee

Name of Expense	Brief Description
12b-1 Fee	How financial advisors or institutions get paid on a trail basis
Expense Ratio	The total of all of the fees mentioned above

There have been extensive studies on expenses by Vanguard and Morningstar and other firms. The BEST performing mutual funds over time have consistently had expense ratios lower than 1%.

I do think some people can take this TOO far. They insist they have to have a fund with an expense ratio of 0.2%, versus another fund with an expense ratio of 0.3%.

I disagree with this premise, because it depends on the funds, how it is constructed, and how it fits into your portfolio.

For example, the Dow Jones Industrial Average (DJIA) and the S&P 500 are two different indices. The DJIA has 30 components and the S&P has 500 components.

The S&P 500 is market-cap weighted. This means that Apple is held at a higher percentage than Microsoft.

However, there's also an investment where the S&P 500 can be equal weighted. This means that Apple is held at an equal percentage to Microsoft.

Same 500 stocks—just held at different percentages.

The market-cap weighted version has a very, *very* small expense ratio, close to 0.10%, while the equal weighted version has a larger expense ratio of 0.50%.

0.50% is still, relatively, a very cheap expense ratio, compared to many high-priced mutual funds.

Despite the higher expenses, the equal weighted version has <u>consistently outperformed</u> the market-cap version. *Note: Not always, but consistently! Also, while I am sure you would like for me to include proof here.*

Unfortunately, I cannot list specific securities here in the book for the whole public due to securities regulations.

Simply, search for "equal weighted S&P 500 ETF" and "market cap weighted S&P 500 ETF" in your favorite web browser and I'm sure you can stumble across what I've seen.

All in all, expenses are very important, but they aren't everything.

The 3 Big Problems with Mutual Funds. There are three big problems with mutual funds—particularly actively managed mutual funds. Index funds, perhaps less so. Although, these same issues can occur there too.

What's the difference you say?

Index funds are mutual funds that try to replicate an index. For example, there are index funds that try and replicate the S&P 500 or the Dow Jones Industrial Average.

In contrast, actively managed mutual funds are trying to BEAT a given index by picking better performing stocks- on both the upside and the downside.

We'll get more to that research later in the text.

Anyhow, there are three big problems that I have with mutual funds...

<u>Problem# 1:</u> You won't know what price you will get when you click the sell button. One of the biggest issues with mutual funds is that they cannot sell at any given moment during the trading day.

They only change their price ONE TIME—after trading hours.

If you want to lock in selling at a really high intra-day price, or buy at a really low intra-day price, you can't do it!

You have to wait until the end of the trading day.

I personally find that really annoying. There are usually a few days a year when volatility rules and we can take advantage of temporary market blips.

Problem# 2: Mutual funds (particularly active funds) are historically horrible at tax management, and many are very tax inefficient. We call this phenomenon "phantom capital gains."

Phantom capital gains can haunt your taxes even when you've had a losing year AND held the fund less than a year.

Check out *The Freedom Formula for Physicians* (www.doctorfreedombook.com) and the "Tax Reduction Prescription" chapter for more specifics on how this dangerous phenomenon occurs.

Problem# 3: Transferability is a huge issue with many proprietary funds. Some companies produce their own products.

Not only are these products very expensive, but often you cannot transfer them directly out! You have to liquidate them at the holding company, first.

HOW ETFS CAN KEEP YOU OUT OF TROUBLE

In order to combat all three of these issues, the Exchange Traded Fund (ETF) was created.

Investopedia notes, *"The first real attempt at something like an ETF was the launch of Index Participation Shares for the S&P 500 in 1989. Unfortunately, while there was quite a bit of investor interest, the Chicago Mercantile Exchange sued to stop them and the advent of true ETFs had to wait a bit.*

"It wasn't a long wait, however; the first ETF began trading in January of 1993. The S&P 500 Depository Receipt (called the SPDR or "spider" for short) was the first of its kind and is still one of the most actively-traded ETFs today. Although the first ETF launched in 1993, it took 15 more years to see the first actively-managed ETF to reach the market."

Here we are, nearly 20 years after the launch of SPY in 1993, and ETFs are overrunning mutual funds. From only 1 ETF to now over 1,000 ETFs!

4 ADVANTAGES OF ETFS OVER MUTUAL FUNDS

ETFs have four specific advantages over mutual funds.

1. <u>Liquidity.</u> ETFs can be bought or sold at any minute during the trading day.
2. <u>Tax efficiency.</u> ETFs buy and sell securities via baskets. Baskets are large chunks of securities like stocks and bonds at the same time. This allows ETFs to avoid the taxation issues that mutual funds run into.
3. <u>Lower Costs.</u> On average, ETFs have much lower expense ratios than mutual funds. They do not have the same trading costs or 12b-1 fees that often drag down the performance of mutual funds, particularly actively managed mutual funds.

4. <u>Transferability.</u> Because they are traded like a stock on the exchanges, they can be transferred in and out of any brokerage account.

SHOCKING ETF CAVEATS TO KEEP IN MIND

ETFs have changed drastically since their dawn in 1993.

They now come in all shapes, colors, and flavors.

You can buy ETFs invested in gold, silver, hospitals, real estate, pipelines, and all kinds of wacky stuff.

<u>The wackier you get—the more expensive, and less liquid the ETF is.</u>

Basically, this means you can get screwed and buy or sell at a horrible price, due to a lack of trading on a very unusual security.

Some of these things only trade a few hundred shares a day.

<u>Guess what else has been happening?</u> The active mutual fund managers who were losing shareholders are now flocking to ETFs. Today, there are BOTH active ETFs and passive ETFs. Passive ETFs are where it all started.

Passive ETFs are home.

<u>Just follow Dorothy's example and repeat after me . . .</u>

There's no place like home. There's no place like home. There's no place like home.

Also, be very aware of the tax structure of the ETF. If they are buying and selling FUTURES rather than STOCKS and BONDS, they may issue a K-1 tax form instead of a 1099 tax form. There are also master limited partnerships and exchange traded notes (ETNs) to avoid some of the K-1 issues.

One piece of advice, you guys: **KEEP INVESTING SIMPLE**.

Don't make it too complex.

To be completely honest with you, I've been guilty of this. I've used some of these "weird" funds. Some turned out fantastically awesome; others really, really hurt and took a few years for my portfolio to recover and regain what was lost.

There are a lot of interesting ideas that look intellectually interesting and "should" work, but when the rubber hits the road, you can become a guinea pig in someone else's grand experiment.

Don't be an experiment—keep it simple!

THE BEST INVESTMENT DECISION I MADE

As I mentioned on the previous page, I have been guilty from time to time of choosing some weird stuff. I've had a core of U.S. equity index funds, but I get a bit bored and ADD kicks in, and I decide to try something a little different.

As I reflect back on my career and the best decisions I've made, without a doubt, one of the best things I did was to buy small-cap stock indices in 2009.

THE BEST INVESTMENT DECISION I MADE, CONTINUED

The market had cratered. Hundreds of thousands of jobs were being shed every month. My revenue had plummeted with the market. It was a scary, scary time. I remember many times being filled with anxiety, wondering what the next day was going to bring.

Warren Buffett once said, "Be fearful when others are greedy and greedy when others are fearful."

In the face of that fear, I purchased the Russell 2000 Index. Rather than buying an individual stock, I purchased a holding that couldn't possibly go bankrupt.

Since that time, it has more than doubled and has been the best investment I ever made. Remember, be greedy when others are fearful.

Have a specific strategy. Make sure it is repeatable and could be replicated by someone else, time after time after time.

Stick to the MOST liquid, passive, traditional pieces of the pie—large-caps, mid-caps, small-caps, maybe even international stocks, high-quality bonds, low-quality bonds, and maybe international bonds.

Learn from my mistakes. I'm telling you . . . Keep it simple!

You may be sitting here thinking… that doesn't sound so simple. Dave- you just mentioned like 10 different potential types of investments!

When I say keep it simple, I am advocating staying away from hedge funds.

I am advocating staying away from investments that are incredibly micro-specific. For example, don't invest in gold miners or biotech indices or underwater basket weaving stocks. Those are all sub-sectors which take on a lot more risk than the index or even the bigger sector.

I am advocating staying away from investments that are privately held and pay huge commissions to advisors that invest in assets like oil wells, real estate, wine companies, equipment leasing, and other such assets.

I am advocating investing in bigger, broader indices for the vast majority of your hard earned money that is invested in your portfolios.

HERE COME IDEAS FOR MUTUAL FUNDS AND ETFS

Okay, so we've laid out the differences between mutual funds and ETFs.

Let me lay out my philosophy for the use of each.

Use ETFs for Stock Allocations. There is no better use, in my opinion, for ETFs. The tax features, the liquidity, and the long-term performance of the indices reinforce my opinion.

According to the 2013 version of the S&P Scorecard, 85.95% of active managers LOST to the S&P 500 (large caps) over a 3-year period, and 79.46% over a 5-year period.

83.63% of active managers LOST to the S&P 400 mid-cap index over 3 years, and 82.88% over 5 years.

With active managers' average expense ratio pushing 1%, it's awfully hard to beat close to 0% in such a liquid environment.

You'd have an 80% chance of picking the wrong active manager!

On top of that, when we focus on wanting to sell at a certain price, bonds have been quite sleepy and boring. They don't move much in a day.

A bond that is intermediate (meaning 3 to 10 in duration- see our earlier chapters for duration) and is investment-grade (again see the 'Why Bonds?" chapters) can move from 0.5% to -0.5% in a day. As a matter of fact, most days it is closer to 0.3% to -0.3%.

Consider with this in mind that a 1% price change in an intermediate bond ETF is VERY volatile for a bond position, but not so much when compared to a 5% swing in a stock ETF on a VERY volatile day for stocks.

Also, tax efficiency is extremely important for equity funds. At the beginning of a bull run, it is routine for stocks to run away with 50% gains.

Intermediate, investment grade bonds don't usually have those kinds of gains. Again, very sleepy and boring. Steady as she goes in comparison to stocks.

Remember—consider the use of ETFs for stocks wherever you can!

There may be some places that they aren't available—for example, your 401k or 403b. In that case, you may want to consider the use of index mutual funds in their place, if available.

USE MUTUAL FUNDS AND ETFS
FOR BOND ALLOCATIONS.

Now, in comparison, bonds have been much more friendly to active managers—particularly in the intermediate space.

The major index that is often used there is the Barclays Aggregate Bond Index.

In the 2013 S&P Score Card, the index LOST to active managers. Passive, index investing won over three years only 38.43% of the time, and over five years only 40%.

Also, globally, the Barclays Global Aggregate Index continued the trend. It won 50.43% over 3 years and 48.78% over 5 years.

I have several reasons why active bond managers are still useful:

1. **Tax efficiency.** Tax efficiency isn't important, due to bonds being INCOME vehicles rather than CAPITAL GAIN vehicles. (See Chapter 21)
2. **ETF Liquidity Not Useful.** There are small price swings in most bond ETFs that render the minute-by-minute, second-by-second pricing not as important.
3. **Illiquidity.** The bond market is extremely ILLIQUID in many parts. There are regular dislocations.

Let me tell you more about the problem with the indices, and the opportunities active bond managers have.

Consider this . . .

Every bond that is issued is DIFFERENT. They have a different interest rate, a different maturity date, and often different provisions and different issuers.

Proctor & Gamble stock is Proctor & Gamble stock. There aren't 2, 3, 4, or 1,000 of them outstanding.

<u>Proctor & Gamble can have 1,000 or more different types of bonds outstanding at any time.</u>

Individual bonds are historically purchased by financial institutions, like pension funds, banks, or insurance companies, to match their liabilities—that are completely different from their assets.

For example, a pension fund may need to meet their lump-sum payouts over the next 20 years. Let's say they have 10% of their group retiring in 5 years and another 20% in 10 years. They may wish to purchase bonds to match those liabilities.

Consider that insurance companies, pension funds, and banks could all desire different maturities for different reasons.

This insurance company may pick one type of bond, and a bank may pick another type of bond, for various reasons.

<u>If they want to sell the bond, it can be very difficult, because they have to sell the bond to someone who wants THAT SPECIFIC BOND compared to the other 999 types of bonds from that same issuer out there.</u>

This leads to a lack of liquidity.

The lack of liquidity of bonds means that the price that BUYERS pay can be significantly different from the price that SELLERS receive.

We call this the bid-ask spread.

Active bond managers take advantage of the bid-ask spread all day long.

They provide liquidity and take it away. The more money they manage, the more power and liquidity they can provide.

The ball is often in their court, and it helps them regularly out-game and out-perform the bond indices.

The bottom line is this . . .

ETFs are fine in the bond portion of the portfolio, but definitely consider active managers of mutual funds as well.

You run a very good chance of outperforming the index in picking a smart, deft bond manager.

Exception: U.S. Treasuries are extremely liquid, due to the demand. There are billions and billions and billions of dollars' worth of bonds floating out there, being bought and sold every day, every month, and every year. The majority of bond managers specializing in the highest quality, liquid bonds have a very, very difficult time beating the index. The statistics bear this out.

If you want U.S. Treasury exposure, consider the ETF and perhaps not the mutual fund.

FINAL THOUGHTS

The investment world has changed a ton over the last few decades. The financial markets (while not perfect) are far more efficient than ever before.

Remember, consider using ETFs for stock allocation. They are cheaper and, on average, give you more bang for your buck.

On the other hand, in my opinion, mutual funds do have a place in your portfolio, and the bond portion of your allocation is a wonderful place to utilize at least a few mutual funds.

When you do use mutual funds, I advocate no-load mutual funds since you can avoid sky-high commissions.

However, if you do use an active manager, make sure you do your due diligence and understand how the manager has outperformed the index, and how they would propose to do so in the future.

The power is in your hands! You are now armed with some basic knowledge to go out and talk to any professional intelligently.

You can do this!

Next, let's cover more about stocks, and break down each component of a typical "moderate" portfolio.

RESOURCES MENTIONED

The Freedom Formula for Physicians:
www.doctorfreedombook.com

The Russell 2000 Index:
http://www.ftserussell.com/

Investopedia:
http://www.investopedia.com/

S&P Scorecards:
http://us.spindices.com/

CHAPTER 23

THE TYPICAL "MODERATE" PORTFOLIO

By Dave Denniston, CFA

B efore we move on, let me tell you a Chinese proverb.

Zhuang Zi was a brilliant philosopher and strategist who lived in ancient China. His abilities were many, and a number of several rulers sought his services. One of them, King Wei, sent his courtiers out to Zhuang Zi's pastoral home to invite him to come to Wei's court and be the leader's chief counselor. They found him there fishing by the river bank.

Seeing his poor situation, they thought Zhuang Zi would jump at the chance for status and reward.

Yet when they made their proposal to him, he said, "Once upon a time there was a sacred turtle, which was happy living his life in the mud.

"Yet, because he was sacred, the king's men found him, took him to the royal palace, killed him and used his shell to foresee the future. Now tell me, would that turtle prefer to have given up his life to be honored at the palace, or would he rather be alive and enjoying himself in the mud?"

The courtiers responded that, of course, the turtle would be happier in the mud.

To which Zhuang Zi replied, "And so you have my answer. Go home and let me be a happy turtle here in the mud."

Source: http://www.fastcompany.com/1809457/3-timeless-parables-regaining-perspective

There are a lot of courtiers who will try to get you to move out of the mud. Their reasons may sound intriguing, or even logical. They may have performance that blows your mind and gets you drooling.

However, let's learn more about the mud. Maybe, just maybe, it isn't so bad . . .

We've broken down bonds quite a bit already. We've talked about risk tolerance and a lot of big picture portfolio stuff. However, it's time to break it down!

Let's dissect and analyze each component of stocks in the "typical" moderate portfolio.

Morningstar Category	% Allocation
Large-Cap Blend	15%
Mid-Cap Blend	15%
Small-Cap Blend	15%
International	15%
Various Bonds	40%

We're going to ask the question . . .

Why should we have mid-caps in our portfolio?
Why should we have small-cap stocks?
Should we have international stocks in the portfolio?

WHAT IS ALL THIS CAP STUFF? CAP OFF!

First, let's have a chat about caps. Cap is short for capitalization.

There's a really easy formula to help understand this word.

Price x Shares = Cap

If you take a stock's price and multiply it times the total shares that are hanging out in the world, this gives you capitalization.

For example, let's say Microsoft is worth $50/share and there's a total of 8 billion shares outstanding—that is equal to a $400 billion capitalization.

In comparison, GoPro is worth $20/share and there's a total of 200 million shares outstanding—that is equal to a $4 billion capitalization.

As you can imagine with the formula, price itself is just one component. A $500 stock and a $50 stock can have the same capitalization, assuming that the $50 stock has 10x the number of shares outstanding as the $500 stock.

A general frame of reference is that large cap is $10 billion in capitalization or above, mid cap is $3 billion to $10 billion, and small cap is below $3 billion.

Generally, large caps are the indices that most folks are aware of. For example, the Dow Jones Industrial Average is commonly known.

The Dow is composed of 30 stocks that are HUGE. Many of them are $100 billion or greater in capitalization. You probably know most of the names in the index.

You also are probably familiar with the S&P 500. These are 500 stocks. The ones that have the greatest weighting, you'd know really well—names like Apple, Microsoft, and Exxon Mobil. However, the bottom 100 are names that probably both you and I aren't familiar with.

That's because the S&P 500 has BOTH large-cap and mid-cap stocks.

You may be wondering . . . why do I need to have mid-caps? Couldn't I just put all my money in the S&P 500 which has some mid-caps in it already?

WHY YOU NEED MORE MID-CAPS

As a matter of fact, one of the frustrating things in 2014 and 2015 was how large-cap U.S. stocks did so incredibly well and left everything else in the dust.

Diversification didn't work! Mid-caps, small-caps, international stocks, and emerging market stocks all lost to large-caps by a big margin.

From time to time, I heard the question, "Well, why don't we put all of our money there in the DOW or S&P?"

To be honest, I was a bit frustrated, myself. I was under the impression that, in a good year, mid-caps and small-caps had done much better than large-caps. Why weren't they up double digits as well in 2014?

On the other hand, since mid-caps and small-caps are generally considered riskier, they should do worse in down markets.

I decided to challenge my assumption and see what I could find. Is this assumption true, or should I re-examine the hypothesis?

In his book, *What Works on Wall Street*, James O'Shaugnessy breaks it down for us.

I read through the book, as I do with most things, and I was pretty skeptical of the findings and research. After all, this is a mutual fund / separate account manager, and there's a conflict of interest in him pointing out that his way was the right way.

One of the biggest aspects I took away from this book was the power of mid-caps. He claimed that mid-caps were more powerful on the upside and protected better on the downside.

This went against my grain of thinking. Weren't mid-caps supposed to be more risky? Shouldn't they have worse downside?

His research went back to the 1950's. I didn't want to go back that far to test his hypothesis because the data is difficult to validate.

Instead, I looked to the era when exchange traded funds started- in 1998. This data was easy to validate on multiple platforms and through multiple methods.

My team and I looked all the way back to 1998, and looked into mid-caps to see what was the high point for the year, the low point for the year, and the year-ending performance.

Name of Index	Average	Number of Years
S&P 500	-11.55%	9
S&P 400 MidCap	-11.78%	8

*Data derived from Morningstar.com & VectorVest

The Low Point. I was surprised by the results. Here's what I found:

If you compared the S&P 500, a large-cap index, versus the S&P 400 Mid Cap index, the low point was nearly equal.

On this table, we're looking at the average low point. They are nearly identical—less than 0.5% apart! As far as I am concerned, it's negligible. (Note: 2012 and 2013 were the same—zero—and thus not counted for these purposes.)

In the far right-hand column, you'll see that we measured how many times each index had the lowest low point of the year between the two of them.

By the way, there are a couple of years where the low point is exactly the same, and so we call it a draw.

What isn't on this table, but is even more fascinating, is what happened in the last two downturns. In 2008 and 2009, large caps did have the edge by 1% or 2%, but in 2000 to 2002, as the tech bubble burst, mid-caps handily beat large caps on the downside—by more than 4% every year in all 3 of the years!

It could be possible in a really mucky year (but not guaranteed) that mid-caps could do better.

The High Point. Okay, we know the downside, but what about the upside? What about the highest point of the year?

Check out this table.

Name of Index	Average	Number of Years
S&P 500	13.09%	4
S&P 400 MidCap	17.09%	15

*Data derived from Morningstar.com & VectorVest

Mid-caps had a significant edge on the upside—almost exactly 4%! Mid-caps were well on their way to an average 20% high point.

Over 18 years, mid-caps beat 'em out, 15 out of 19 times. That's also big! It's not just 1 or 2 years as outliers—it's consistent.

What isn't in this table is that the recovery years have been particularly robust. For example, in 2009, large caps were up about 25%, and mid-caps were up 38%; and in 2003, large caps were up 26%, versus 35% for mid-caps.

So, not only are mid-caps about equal, if not possibly better on the downside, they kill it on the upside!

The Year-End. Then, finally the year-end table looks very similar to the high-point table.

Name of Index	Average	Number of Years
S&P 500	6.09%	6
S&P 400 MidCap	10.03%	13

*Data derived from Morningstar.com & VectorVest

See how mid-caps have consistently outperformed large caps?

The year-ending average is getting close to double!

All right, let's bring this all together.

While 2014 and 2015 had large caps beating nearly every single asset category, mid-caps have consistently been a good piece of the puzzle.

Any one year can have its fluctuations.

In brutal years, it's about a 50/50 shot that large caps can outperform during the low points in the markets.

What I like about mid-caps is that they can and usually do stay fairly close to large caps, and give us a great chance to outperform the large cap index.

The question is: Why?

If you look at many mid-caps companies, they are the up-and-comers. They are the disruptors. I believe that they've transitioned from a small-cap to a mid-cap and the company is growing, growing, growing. Eventually, they become a large cap and are included in the S&P 500 as a larger and larger weight.

Meanwhile, some of the large cap companies are so big that for them to grow can be very difficult. As a recovery takes hold, the mid-cap disruptors are growing into new areas and can recover quicker.

It's very interesting! <u>Make sure to overweight mid-caps!</u>

BUT, WHAT ABOUT SMALL-CAPS?

All righty, so we know that we need more mid-caps—those companies that are valued from about 3 billion to 10 billion.

We looked all the way back to 2001 (again when the ETF started to validate the hypothesis in data that anyone could double check) through 2015 and looked to see what was the high point for the year, the low point for the year, and the year-ending performance.

I wasn't too surprised by the results on the downside, but I was surprised by the upside.

Here's what I found:

Low Point. If you compared the S&P 500, a large cap index, versus the Russell 2000 index, the low point was significantly different.

Name of Index	Average	Number of Years
S&P 500	-12.58%	11
Russell 2000	-14.36%	3

*Data derived from Morningstar.com & VectorVest

On this table, we're looking at the average low point. (Note: 2013 had no down point from the start of the year for both the Russell 2000 and the S&P)

If you look on a year-in-year-out basis, there's no doubt that the S&P has done consistently better—11 times out of 14. Whereas, the Russell 2000 consistently has a lower low point every single year.

Some years did vary more than others—the brutal years are usually *more* brutal. For example, in 2011, the lowest low point was -12% for the S&P versus -22% for IWM.

It can be significantly different, and in a bad year, it could hurt more and be more volatile.

High Point. Okay, we know the downside, but what about the upside? What about the highest point of the year?

Check out this table.

Name of Index	Average	Number of Years
S&P 500	12.42%	3
Russell 2000	16.68%	12

*Data derived from Morningstar.com & VectorVest

Small-caps had a significant edge on the upside—almost 4%! Small-caps were well on their way to an average 20% high-point.

Over 15 years, small-caps beat 'em out 12 out of 15 times. That's also big! It's not just 1 or 2 years as outliers—it's consistent.

What isn't in this table is that the recovery years have been particularly robust. For example, in 2003, large caps were up about 25% and small-caps were up 48%; and in 2010, large caps were up 14%, versus 27% for small-caps.

Year-End. So, small-caps can be very painful on the downside, but they do amazing well on the upside!

Finally, the year-end table looks very similar to the high-point table.

Name of Index	Average	Number of Years
S&P 500	4.7%	4
Russell 2000	7.99%	11

*Data derived from Morningstar.com & VectorVest

See how small-caps have consistently outperformed large caps by the end of the year?

The year-ending average is getting close to double!

All right, let's bring this all together.

While 2014 had large caps beating nearly every single asset category, small-caps have also consistently been a good piece of the puzzle.

Any one year can have its fluctuations, and during cruddy years, it's about an 80% shot that large caps can outperform during the low points in the markets.

This is why, when I use tactical asset allocation, I will minimize the small-cap exposure.

However, what I like about small-caps is that they can drastically outperform the large-cap index on the upside.

There are no guarantees, but the data is pretty powerful—particularly robust in early recovery years.

Pretty cool stuff!

Note: Our more aggressive risk tolerance folks may want to allocate MORE to small-caps while perhaps less risk tolerant folks should have LESS small-cap exposure.

WHAT ABOUT INTERNATIONAL INVESTMENTS? THE CONUNDRUM

We've talked about the place of mid-caps and small-caps and why they are still good to have.

Now let's explore international stocks.

This can be confusing, because it depends on what country, what region, or how you put which one in what bucket.

So, let's ignore individual countries, and instead focus on big classifications of international stocks.

There are four main areas to be aware of: global, international developed, emerging markets, and frontier markets.

Global is a bit confusing, because you think of other countries, but in fact, <u>most indices that are categorized as global include about 50% U.S. companies</u>.

If you have a global fund or a global ETF, it is likely dominated by U.S. companies. However, you can still get a taste of international companies.

In comparison, international developed indices don't have U.S. companies, but they focus instead on developed areas. For example, they contain stocks based mostly in Japan and Western Europe, with a couple other countries thrown in here and there.

Then, emerging markets—or EM as we like to call them—are really where most of the population of the world is contained, and their economies have been up and coming; the majority of those indices are filled with B-R-I-C . . . BRIC for Brazil, Russia, India, and China.

As you can imagine, emerging markets have been beaten up pretty badly with China's economic rockiness over the last few years.

Lastly, frontier markets are the up-and-comers stretching to make it into EM—think of countries like Vietnam, Saudi Arabia, South Africa, Argentina, and Kenya.

What's frustrating about international investing, even in a broad, diversified play, is that performance can be so much better than domestic stock at times, but then on the other hand international investing can really lag behind domestic sometimes.

Check this out first.

S&P 500 versus a Global Index: Remember that 50% of the global index is U.S. companies—so we should expect some similarities.

ANNUAL PERFORMANCE OF S&P 500 VS. FTSE GLOBAL

Index	2008	2009	2010	2011
S&P 500	-36.8%	26.37%	15.06%	1.89%
FTSE Global Allcap	-42.2%	37.5%	14.8%	-7.6%

Source for all tables: Morningstar.com & VectorVest.com

Index	2012	2013	2014	2015
S&P 500	15.99%	32.31%	13.46%	1.25%
FTSE Global Allcap	17.2%	23.9%	4.5%	-1.7%

So, definitely, there are a lot of similarities here, but a few differences, too. 2008 was comparatively poor for the global index, but then it recovered more in 2009. It lost a bit more in 2011, but recovered, mostly, in 2012.

However, in 2013 and 2014, the S&P outperformed it in both years by nearly double digits. So, domestic stocks are doing better lately, but since global stocks have the S&P 500 in them—it can't stray too crazy far.

I think you can see where I'm going with this.

Next, let's look at EAFE, a developed markets index, versus the S&P.

ANNUAL PERFORMANCE OF S&P 500 VS. EAFE

Index	2008	2009	2010	2011
S&P 500	-36.8%	26.37%	15.06%	1.89%
EAFE	-43.4%	31.78%	7.75%	-12.1%

Index	2012	2013	2014	2015
S&P 500	15.99%	32.31%	13.46%	1.25%
EAFE	17.32%	22.78%	-4.90%	-0.81%

Here's where we start to see some significant differences, you guys. 2008 and 2009 look pretty similar to the Global Index. But then 2011 until today has wider swings—the lows are lower—for example, -12% in 2011 and -5% in 2014.

In 2015, the EAFE stayed relatively close, despite the Greek drama over the summer.

As of the time of this writing, EAFE is lagging behind once again—as "Brexit" has become a word we are all adding to our dictionaries.

Index	10-Year Average
S&P 500	7.77%
EAFE	1.98%

Here's where the rubber meets the road: the 10-year average.

The S&P 500 pulled in a respectable return that was nearly 8%.

In stark contrast, EAFE is LESS than half the return of the S&P 500. Heck, it's only about 1/3 of the return of the S&P. You would have lost out on MORE than 4% every year by being invested internationally rather than domestically.

Ouch!

All right, let's move onto emerging markets.

ANNUAL PERFORMANCE OF S&P 500 VS. FTSE EMERGING MARKETS

Index	2006	2007	2008	2009	2010
S&P 500	15.85	5.14	-36.8%	26.4%	15.06%
FTSE EM	33.1%	39.7%	-52.9%	82.6%	19.8%

Index	2011	2012	2013	2014	2015
S&P 500	1.89%	15.99%	32.31%	13.46%	1.25%
FTSE EM	-19.0%	17.9%	-3.5%	1.6%	-15.2%

This is where things get a big crazy.

I squeezed in a couple of earlier years to explain the allure of EM. Check 2006 and 2007. EM was flying—killing U.S. equities. Then, they dropped like a rock. Way worse—look at 2008.

Then, EM stocks took off like a rocket in 2009—tripling what the S&P did, followed by another great year in 2010, and 2011 was a really horrible year for EM—but NOT for U.S. stocks, followed by a make-up year in 2012.

Definitely- very volatile!

Since then, it's been very frustrating to be in EM stocks. U.S. equities zoomed up 30% in 2013, and yet, the GDP of India and China have continued to climb over this same period.

Whether it is due to accounting fraud or government manipulation or bad actors all around, that growth has not helped their stock markets.

This same trend continued in 2014 and 2015. Emerging markets equities have been at the bottom of the barrel. As of the time of this writing in August 2016, emerging markets are finally beating U.S. equities—but have a ton of room before they can catch up.

Index	10-Year Average
S&P 500	7.77%
FTSE Emerging Mkts	3.87%

Check out the 10-year averages—EM actually beat EAFE, but lost ground to the S&P 500, not quite as badly as EAFE, but still by nearly 3% per year—or 30% over the last 10 years, which we know now is mainly because of the last couple of years.

2014, EM went nowhere, followed by a really atrocious year again in 2015.

To be honest, I've had money in EM, and it's angered & frustrated me tremendously.

However, on the other hand, it seems due for a bounce in a big way if you look at the past.

Ah, but will it?

Will emerging market stocks catch up to U.S. stocks?

That's the conundrum.

As you can imagine, emerging markets have held back more risky portfolios over the last couple of years.

If you re-ran the model portfolios we listed earlier by risk tolerance and took out emerging markets, it would drastically affect the higher risk tolerance portfolios. You would have added another 1% up to 3% in annualized returns.

Will the past repeat, or will it rhyme with the future?

Are the correlations permanently disconnected?

Each of us will need to decide if we want to take on that risk.

The reward could very well be worth it.

However, you'll have to grit your teeth in frustration if it doesn't work out.

Personally, I'm planning on holding on to this asset class for the next couple of years. I'm a believer . . . but I might lose all of my hair in the process, and pop a few pills while I am at it!

Anyhow, lastly, let's move on to Frontier Markets.

ANNUAL PERFORMANCE OF
S&P 500 VS. BNY MELLON NEW FRONTIER INDEX

Index	2009	2010	2011	2012	2013	2014	2015
S&P 500	26.4%	15.06%	1.89%	15.99%	32.31%	13.46%	1.25%
BNY Mellon New Frontier Index	33.9%	33.9%	-22.4%	12.3%	-14.6%	-11.8%	-21.44%

What we see here with frontier markets is very similar to emerging markets. It's been even more volatile, the last few years: negative double digits in 2013, 2014, and again in 2015.

What's interesting to me was that 2009 wasn't nearly as good as emerging markets.

So, there's a ton of opportunity here—but even more risk. There's more geo-political risk and instability in a lot of these places.

I've personally never invested there, and don't intend to.

Frontier markets is just too much risk for me! However, if you have a super high aggressive risk tolerance, go for it!

Just remember: There could be a huge disconnect between frontier markets and domestic stocks.

Frontier markets could lead you to drastically underperform or overperform domestic equities—even more so than emerging markets stocks.

FINAL THOUGHTS ON "THE MODERATE PORTFOLIO"

All right, so where does all of this lead us?

We really have to ask ourselves: Do we really want to have international stocks in our portfolios?

<u>How would we feel if having international stocks causes us to underperform when domestic stocks are up?</u>

One could make the case that with the Fed raising rates, and countries like Japan and the Eurozone adding stimulus—that stimulus is equal to better stock markets.

We could add a global index in, to put a foot in the door, but mostly stay domestic. We could add in developed markets, believing the stimulus story.

We could add in emerging markets, thinking that they'll finally bounce back—or even frontier markets.

Personally, I think Japan and the Eurozone are going to have a very hard time growing their economy, and while exports might do better in the short term, due to the dollar appreciating, I don't like the long-term picture.

Emerging markets continue to be very attractive—I think there's a lot of potential there—and they are moving toward increasing the power of the consumer in China and India, and these countries are implementing many reforms. I like the long-term picture there. It just comes with a lot of risk over the short term—but also a very large potential reward.

I think staying focused on U.S. markets keeps managing a portfolio simpler and easier to understand. After all, how many of us track how EAFE is doing, or how emerging markets are doing?

On the other hand, many opportunities could be missed if you are not invested in emerging markets.

Overall, I'm personally committed to emerging markets in the next few years, until they've bounced back significantly. The potential is too much to give up.

However, I'm coming to realize that it can lead me to stray too far away from what happens at home, and so over time, I am planning to keep some of that index in riskier models- but will make sure to eliminate or lower it for others.

I'm not going to do it, though, until we get that growth back. To sell now and never buy back would be foolish, in my opinion, when they are so relatively low.

RESOURCES MENTIONED

What Works On Wall Street by James O'Shaughnessy:
www.doctorfreedombook.com

The FTSE Global All Cap Index & The FTSE Emerging Markets Index:
http://www.ftserussell.com/

VectorVest- Software for Historical Tracking & Paper Portfolios:
http://www.vectorvest.com/

Morningstar: Ranking and Research of Mutual Funds, ETFs, and
Individual Stocks:
www.morningstar.com

CHAPTER 24

THE WEIRD STUFF LURKING
IN YOUR PORTFOLIO

Imagine sitting there. You're opening up your monthly statement. The ripping of the envelope fills your ears as you anxiously await to find out how you are doing.

You haven't paid attention to your investments lately, but you know the markets have been doing all right, as you've talked with your colleagues.

You snatch the statement out of the envelope. The envelope drifts to the floor.

As you scan the statement, your jaw drops. *You've lost money.*

You can't believe it.

How could you lose money while the markets have been going up? What the heck are you invested in?

My friend, many physicians have been in your situation.

Let's find out why.

ALTERNATIVE ASSET CLASSES

As mentioned earlier, there are a host of mutual funds and ETFs that have proliferated like wild rabbits.

Many are simply hedge funds in disguise. Super high fees and sub-par returns.

The huge majority of these SUCK.

They lure us away with promises of not investing in the stock market, and using alternative asset classes to hedge against risk.

This often means the managers are using derivatives like forward and futures that can be EXTREMELY EXPENSIVE.

Futures and forwards are what we call 'derivatives'. To make a long story short, they are contracts that usually bet on a short-term direction of an investment.

These could be stocks, bonds, commodities like oil, and currencies.

There is a cost to buying futures every time you enter into a new contract. Since many of them expire every one, two, or three months, you are constantly rolling over the contracts by getting new ones.

This can be very expensive.

Morningstar throws around a few categories like market-neutral or managed futures, because these vehicles are hard to understand.

In Morningstar's print magazine of August / September 2015, they noted, "Long-short equity funds, for example, have grown almost five-fold since 2008, as investors were attracted to the category's relatively superior performance during the financial crisis—the category dropped only 15.4% while the market had a 37% pull back. <u>But the funds, as expected, haven't kept up as well during the market rallies.</u> *They have lagged the market by an astounding 11.5 percentage points per year from 2009 through June, while only beating the Barclays Aggregate Bond Index by 0.2% over the same time period.*"

'Nuff said! Don't do it!

SECTOR ROTATION STRATEGY

Some of us (like me) have a bad habit and get bored with simply buy-and-hold, rebalance.

> ### MESSY MATTERS
>
> One of my earliest jobs in my career was being a paraplanner.
>
> I didn't have enough in assets and income to make the transition yet to being an entrepreneur. So, I did a bunch of other administrative work and support. It's kind of like residency!
>
> My boss was a good guy, and he genuinely cared about his clients. He treated me, as an employee, really well.

MESSY MATTERS, CONTINUED

However, he had a philosophy, at the time, that really made me scratch my head. He wanted to make his clients' situations become so complicated that they NEEDED him.

Well, what did that mean? He put a portion of their investments in complicated products. In particular, managed futures. Often these were private products, and there were restrictions on liquidity.

This meant that if clients wanted to get their hands on the money, they couldn't get it immediately. As a matter of fact, it might take as long as 3 to 6 months. To add insult to injury, these product companies charged hedge-fund-like fees—assets under management of around 2% to 3%—plus they captured 25% of all the profits.

When you looked at their past performance, it was pretty impressive. They protected completely on the downside—even making money when equities were down, as well as making money when equities were up. Yet, the fallacy was that this was back-tested performance. It was based on a model—not actual reality (or when it was much smaller).

Even if it was reality, I firmly believe there's no free lunch. Something that looks too good to be true, that you can't replicate yourself—probably *is* too good to be true. As clients experienced this investment, they became pissed off. It lagged and lagged and lagged the market.

I left nearly 10 years ago, and apparently during the great recession, this investment did a horrible job of protecting, and clients were ticked that they couldn't liquidate it.

We know that long-short and market neutral funds are horrible investments. We know that international funds have been extremely painful to hold and have lagged domestic stocks.

What's a person to do if they want to do something different?

There's nothing wrong with taking more risk on the fringes—if you are willing to gain or lose to the index. Just make sure such strategies are a small part of what you do.

In the next few pages, I am going to reveal to you my newest strategy, my newest system that I started rolling out a few years ago. The results have blown me away in some years and frustrated me in others.

As a matter of fact, I discovered that this new system only needs to be tweaked one time per year. Let me tell you about how I came across this newest strategy.

A few years ago, I was at this point where I couldn't sleep. I was tossing and turning.

I always want to be the best, to do better than my peers. Simply, I wanted to do better. I finally got up in the middle of the night, for a few nights in a row, and discovered some pretty awesome research.

Just so you know: By nature, I'm a contrarian. When people are zigging, I want to zag. If everyone has an Apple product, I'll buy Google. When people are selling, I want to be buying.

I had lunch with a friend and a colleague one day, and we were discussing various strategies—the subject of sectors and sub-sectors came up. The idea was to buy certain sectors rather than buy the whole stock market index and I thought it was a nice idea at the time, but wasn't sure how to apply it.

I was hesitant and thought, "Great, another shiny object— Squirrel!"

As I mentioned earlier, that night I couldn't sleep. I had this thought, I literally felt a surge of inspiration . . .

What if we bought the worst-performing sector of the previous year, buying it on January 1st and selling on December 31st, then bought the next worst-performing sector?

Unfortunately, getting data on ETFs and sectors that have been traded is rather limited. So, the farthest I could go back was 2001 to 2002 depending on the index.

Unlike the mid-cap/small-cap discussion in the previous chapter, there was no book written to double-check the long-term data.

Nonetheless, I was curious to see the results.

I tested and re-tested and had interns run a whole bunch of calculations, and found that by using the major sectors, if you buy what has been beaten up the worst, it can lead to an average annual spread over the S&P 500 by 4% over a period of the last 16 or so years.

This especially worked well in years after a crash—for example, tech in 2002, materials in 2009.

However, due to only holding one sector, it could vary pretty drastically from the market in any given year.

What if rather than one sector, we bought the two worst-performing sectors of the last year? We buy the two worst sectors on January 1st and sell on December 31st. Then, we bought the new two worst-performing sectors and held them for the following year. Then, rinse and repeat.

It was just as good, and more consistent in its returns, hugging the S&P 500 much closer and still exceeding the S&P by 2% a year.

We ran and re-ran scenarios, testing what if, rather than starting in January—starting in February, then March, then April, and so on, holding for an entire 12-month cycle.

In nearly every single cycle, no matter where we start, this system, buying the beat-up sector(s), works!

Full disclaimer: It doesn't work every single year in beating the S&P 500, but it stays relatively close. For example, with buying the 2 beaten-down sectors, it has beaten the index with the overall spread of 2% a year mentioned earlier.

2015 was a horrible year for this strategy, as energy and telecom were purchased. Telecom did just fine, but energy tanked. In 2016, energy was still the worst performing sector and materials replaced telecom. Energy and materials doubled the index's performance in 2016.

This is very similar to the DOGS of the DOW Theory where rather than buying sectors and ETFs, they are buying beaten up individual stocks.

I find there is too much individual company risk in stocks and thus, why I love this new strategy as my "one-off".

So, there's definitely some concentration risk here—but at least you are diversified among many different companies.

Anyhow . . .

I've personally made some mistakes in the past, investing in emerging markets, dollar index, gold, or other "weird" sectors. It's become clearer and clearer to me that clients want to stay closer to the U.S. market. It is what they watch and focus on, and so that's what we need to focus on.

What's so awesome about this strategy is that it sticks to U.S. stocks.

This isn't some weird market-neutral fund, or international investments.

It isn't based on opinions or thoughts. We rotate every year, and this will help us to avoid holding on to losers too long and hoping they catch up, when there could be a better opportunity for the next 12 months.

It won't be perfect, but at least it will be easy to track, easy to pull the trigger, and easy to rotate. It's completely unique, and something that I've never seen someone implement in this manner.

(Hint: This is best for Roth and IRA dollars where we have no capital gain taxes to pay. If you use this with non-qualified money instead, the advantages could be partially or totally erased.)

Can you see the benefits of this strategy? Do you see how we can buy the worst beaten-down sectors and simply rotate one time per year?

How we wouldn't stay tied to one sector for years at a time?

For example, imagine what it would have been like if we'd bought tech in 2003, or financials and materials stocks in 2009.

Then, a year later, we moved on to a new opportunity.

Pretty cool stuff!

If you have any further questions about this particular topic, feel free to contact me.

FINAL THOUGHTS ON INVESTMENTS

I just simply want to rinse and repeat what we emphasized earlier.

Generally, <u>keep investing simple.</u>

If you want to explore something different, consider the sector rotation strategy that I've described.

DON'T USE HEDGE FUNDS or MUTUAL FUNDS that are wolves in sheep's clothing where they are using futures and forwards.

Take some time to consider the bond portion of your portfolio.

How are you planning (if at all) to hedge against interest rate risk over the next 10 to 20 years?

I haven't (and likely you haven't, as a young physician) seen a cycle of rising interest rates in our lifetime! The last time that interest rates consistently rose was from the early 1960s until 1981.

Are you using ETFs and/or mutual funds?

Are you using international investments now? Will you be using international investments in your mix in the future?

Make sure to avoid those crazy managed futures and market-neutral funds.

If you want to try something a little fun and out of the box, consider the sector rotation strategy!

More questions?

Contact me at dave@doctorfreedompodcast.com

Home Buying

CHAPTER 25

BUYING YOUR FIRST HOME
By Dave Denniston, CFA

I magine . . . Picture this.

You are sitting at your kitchen table, fingers tapping anxiously. You hold your breath as you sign the final piece of the closing documents. Across from you, a crony from the local closing company awaits your signature . . .

Your fingers feel cramped from the countless signatures you've already made. You wonder, *Is this it? Am I finally done?*

You're feeling so confined in your apartment. You're ready to break free and own the ground that you live on. You're ready to own the American dream.

You're feeling like you are wasting your rent dollars every month. Couldn't you be building equity, instead?

You exhale slowly as your pen makes one last squiggle of your barely-recognizable signature to close on your first home.

You've done it . . . you've closed on your first home!

A feeling of relief—mixed with unstoppable excitement and a tinge of fear—washes over you.

Let's take a step back for a moment.

What would that feel like for you?

Close your eyes and imagine what your home would look like.

Picture the outside. The shrubs are bursting with flowers. The sweet smell of freshly cut green grass neatly manicured in your yard is wafting in your nostrils. The leaves of your trees swirling in the wind. The pungent, warm smell of the rainbow-colored flowers and the warmth of the sun heating you from the outside to the depth of your being.

Walk inside your home.

What do you see?

What do you hear?

What do you smell?

Can you imagine the feel of the mahogany hardwood floors, the soaring ceilings above your head?

Can you see yourself relaxing on your comfy leather couch?

Now, that you've pictured your first home . . . are you excited and ready to go?

Awesome!

However, let's slow down for a minute and consider all the dos and don'ts of mortgages.

As a resident, and as a young practicing physician, there are a lot of traps and temptations that doctors can get into.

Our goal is to keep you from getting ensnared in one of these bad boys, and to have you keep moving eternally in the right direction.

It's unfortunate but true that many physicians are taken advantage of by unscrupulous lenders who prey on their lack of knowledge. You're a rich doc, right? You can afford it!

NO!

Let's journey down the road together and explore what you should be looking for, what the house-hunting habits of millionaire physicians are, and what makes a good mortgage versus a bad mortgage.

MY FIRST HOME

It was longggggg day. My wife and I looked at house after house after house after house. To be honest, they were all blurring together for me. I couldn't really remember one from another.

I was getting a headache. All I knew was . . . they weren't what my wife was looking for.

I was ready to go home, but our realtor and my wife insisted that we go on. I grudgingly agreed to see . . . yes, one more house.

We wove our way into a relatively new neighborhood—it was your typical modern construction—all house and barely any yard—**you could practically reach across and touch your neighbor.**

I was thinking . . . *This isn't promising.*

Promptly, I noticed something weird was going on. It was wayyyy too dark.

Pitch. Black. Dark.

I was like, "Where's a candle when you need one?"

As a matter of fact, I realized that all the houses in the neighborhood were completely dark and all the lights were off.

No street lights. No yellow, shining like a beacon, from any home. No lights of any sort were lit.

We parked in front of our target and discovered that the darkness was due to a power outage!

Now, this house—I have to admit—was different from the rest of the neighborhood. Maybe it was just the darkness, but it actually seemed kind of nice!

The driveway was three times longer than the neighbors', and there was a nice-sized front yard with a small play structure for kids sitting on the street. Just as important, it didn't appear that you could just reach over and touch your neighbor's bedroom window.

However, it was still super dark! There was no way we would go into this place.

I wiped off my brow, thinking, *Whew, good thing . . . I'm ready to call it a night!*

It was crazy dark—I could barely see more than past the realtors' car lights.

Then . . .

. . . my wife dropped the bomb.

"We should go in!" she exclaimed.

Are you freaking kidding me?

There's no lights, woman! I thought.

Giving me a discerning look, the realtor said, "Let me turn on my brights and see what we can see inside."

Cautiously and carefully, we headed into the pitch-black house, bright headlights highlighting our looming shadows in front of us.

However, as we walked in . . . something finally felt right here, a warm feeling expanding in my gut.

This felt different.

From what little I could see, I was impressed. Hardwood floors, a spacious interior, a kitchen island, a walk-in kitchen pantry, and so much more.

Yet, I couldn't see jack squat! We were feeling our way slowly around the house, trying not to trip over objects and smash our faces or bodies into the ground.

My wife exclaimed, "This is our house!"

Okay, she's a little crazy.

I said hesitantly, "Yes, it seems nice . . . but I'd actually like to see it with some lights on."

But you know what?

She was right! This was our house.

And so, shortly after we actually did see the place a few days later, we started negotiations.

HOME-PURCHASING HABITS OF MILLIONAIRE PHYSICIANS

While we don't talk much about negotiating here, there are some basic principles that you should know.

BEFORE even thinking about buying your home, I'd like to point out a few characteristics of millionaire physicians from *The Millionaire Mind* by Thomas Stanley.

Here are the top house-buying habits of millionaire households:

- Buy an older home in an established area (Rarely, if ever, build from scratch)
- Never pay the full asking price for a home (Negotiate)
- Millionaires research the area around the home and comparable sales before they buy it (Do the homework!)
- Millionaires take weeks—even months—to find the right property (Don't make a hasty decision with large purchases)
- They test price sensitivity (Negotiate bargains and discounts)

Consider these characteristics as you shop.

Without necessarily knowing all of these in our mid-20s, my wife and I bumbled around and applied most of these principles.

The main takeaway here, my friends—shop, shop, and shop some more. Don't settle!

Make sure you negotiate!

In addition to some pitfalls we're going to cover in the next chapter, here are a few other principles that I'd like for you to keep in mind.

- Consider the affordability of your prospective home. If you live in California or New York, it may be worth delaying a purchase longer. Whereas, if you are in a small town in the Midwest, it could be fairly easy to afford a home.

- How does it affect your other goals? For example, are you better off becoming consumer-debt-free first?

- Is this going to be your "final" home, or is this merely a step along the way? Are you starting small and then getting your dream place? Or are you hopping right to your dream place?

All right, so let's say that you were pre-qualified for a home, bought an amazing first place, and now you need to line up a mortgage.

What kind should you consider and why?

A TALE OF A GOOD MORTGAGE VERSUS A BAD MORTGAGE

It was June 2007, and the prices of homes in the area were going up, up, and up.

They would never come crashing down.

Money was easy and flowing like a river.

The sun was shining brightly in Seattle, yet I could see a storm of clouds approaching.

A tall, lanky gentleman named James had an easy-going nature. He had stars in his eyes and cash in his wallet. He was a pilot, a flyboy. He'd pick up anything with a thrill. He loved motorcycles and living big.

He had big aspirations, too. He loved him some real estate.

He bought house after house after house. He levered up every one, used five-year adjustable-rate mortgages (ARMs) where he only paid interest. As a matter of fact, he only had to put down a couple of percent as a down payment.

On top of that, he wanted to invest in cash value life insurance policies and borrow "for free" from them. At that time, I was a fledging para-planner and everything in me screamed that this was the wrong thing to do. (Note: A para-planner is like a resident in the financial planning world- you aren't a full practicing financial advisor and you are learning how the business works. It often involves a lot of paper shuffling and case preparation. However, you are also actively engaging with clients and assisting with advice.)

My goal was to be debt-free, not saddled with debts—and then "invest" in a life insurance policy with copper handcuffs that chafe your wrists?

No, thank you!

I have to admit, I wasn't there to see it unfold. However, given his lack of liquidity, I have to imagine that the whole thing collapsed as easily as a house of cards. A weak wind could have blown the whole thing down.

My friends, this was the wrong kind of mortgage
for the wrong kind of situation.

In comparison, consider this situation . . .

A couple in their late 20s were sick and tired of being in their apartment. As a matter of fact, they wanted to start their family and have the house ready to go for the first kiddo.

They had scratched and scrimped away every extra dollar they could.

Luckily, they had already paid off their student debt and now were experiencing more and more cash flow.

Unfortunately, in Seattle, prices were extremely high. They had been exploring various neighborhoods, and settled on a neighborhood near Greenlake.

The homes were older and established. Yet, most of them cost more than $500,000. This meant to have 20% down, they would need about $100,000.

They had saved.

BUT NOT THAT MUCH!

As they started to explore mortgages, they found one at a relatively attractive fixed rate of 5%, and they could put down nearly 10%.

However . . .

They were going to have to pay PMI (private mortgage insurance). For a time, they were willing to bite the bullet and pay more than a 'normal' conventional mortgage. (Note: We cover PMI more on Chapter 27 if you aren't sure what it is)

As a matter of fact, to try and get rid of the PMI, they put an extra few hundred dollars a month toward accelerating the process. Fast-forward a couple of years, and their home had increased in value! On top of that, rates were slightly lower.

This was a wonderful recipe to allow them to refinance their home and no longer have to pay PMI.

They knew this was the home that they wanted to stay in. They had no plans to move elsewhere.

Thus, a new fixed-rate mortgage was a good answer for them, and worked fantastically.

Although . . . one could have argued that a variable rate might have been better.

Let's find out why.

FIXED RATE VERSUS VARIABLE RATE

As an entrepreneur who has been through two severe financial crises in my career, I have to admit that my perspective has been tainted some.

For example, I tend to be very risk averse when helping people make BIG decisions. Why take a chance that could drastically impact you negatively, financially, when you can get a sure thing?

When mortgage brokers talk about variable rates, I get this queasy, uneasy feeling in my stomach. It churns and burns, I wiggle and squirm in my seat as I consider the potential consequences of this situation.

Scenes flash before my eyes of mortgage statements with increasing payments and homes getting foreclosed on, people being forced out of their homes.

So, yes, I certainly have a bias toward fixed-rate mortgages.

It's a sure thing. You know exactly what you are going to pay.

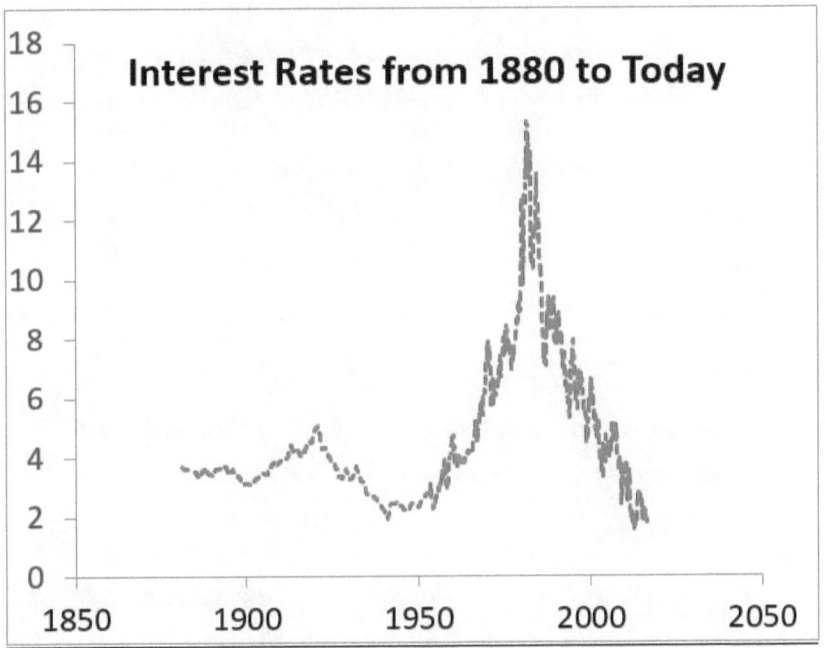

Source: Yale & Robert Shiller
http://www.econ.yale.edu/~shiller/data.htm

In the graph, check out the interest rate cycle. From a long-term perspective, it's pretty clear that rates are going to rise in the future.

There's no question—there's some empirical evidence to support the case for a fixed-rate mortgage.

The question is—how far and how fast will interest rates rise?

One of my previous podcast guests, Josh Mettle, a physician home loan specialist and the author of *Why Physician Home Loans Fail*, has a different bias toward mortgages.

He prefers an adjustable-rate mortgage. I asked him why. Here's what he said.

"This is a great question. I actually have a bias toward adjustable rates, and here is why. I find physicians early in their careers have upward mobility in terms of income.

As such, they quickly find themselves with more cash, less debt, maxed-out retirement, and they seek a nicer home. On average, I find this happens around 5 to 7 years in their home.

So, if we agree that younger physicians move 5 to 7 years on average, think how much money is lost nationally with paying for a fixed rate on years 8 to 30 when they are not in the home.

It's staggering! Play with this question, how many of your clients under the age of 40 have been in the same home for more than 7 years.

I bet you will find that 90% of them have been in their home less than 7 years. Keep record for the next month and see what you come up with."

Josh was then kind enough to send me a case study of an adjustable rate 7/1 ARM (this is a rate locked for 7 years) versus a fixed rate.

As of the date of this writing (July 2016), an ARM had a lower rate—3.625% versus a 30-year fixed rate of 4.125%.

Assuming you had a $400,000 mortgage, this led to monthly savings of $109/month on a payment. That's $1,200 a year!

So, there's no doubt that you could be doing something else with that savings—paying down your non-deductible student debts, or socking money away in a retirement account, or just simply enjoying life more. After all, $1,200 could fly you to Europe or Asia or someplace fun.

The question is this . . .

Do you see it as a distinct possibility that you could "upgrade" to a new home in the next few years?

Or is this the home that you see yourselves staying in for years and years?

Is it a strong possibility that you could relocate to a new gig?

Josh makes a good point here, that perhaps a fixed rate isn't always the answer.

My experience in working with physicians is that if they buy a home as a resident, or right when they transition to practice, then YES, they will end up upgrading to a new home in a new location down the line. An ARM could be a great fit for that physician.

However, by a year or two in transitioning to practice, most physicians (who haven't bought a home yet) have a good idea that this is the place that they want to be (or not).

Once they do buy a home, these physicians tend to stick to what they bought. They aren't buying a condo or a townhome. They are buying the single-family home that they want to be in. In this case, a fixed rate is likely a better bet.

Which of those two scenarios do I prefer?

I prefer the latter. Get planted and know that you know that this is the place you want to be, then buy a home.

I see a lot of physicians stressing out about the homes they bought in residency. They struggle with the decision—*Should I rent it? Should I sell it?*

They gnash their teeth, grinding over the decision. In many cases, as a rental property, they are cash-flow negative. Yuck! It's sucking cash away, every single month.

Now, on top of that, they are saddled with an adjustable-rate mortgage where the rate is likely to go up.

If you are reading this book in the year 2021 or 2022, let's say rates have risen a few percent. Maybe there's a very good chance that rates may not rise much further.

Perhaps right now, adjustable-rate mortgages could be a good fit for many scenarios—because rates could fall back down. So, keep this in mind as the markets change and rates fluctuate. I definitely don't think that fixed rates are a cure-all. There's definitely a time when a variable rate could be much more attractive overall.

Anyhow, your situation could be different! Definitely consult with your advisory team—your tax expert, your financial advisor, and your mortgage broker—to figure out what scenario is best for you.

Ok. We've talked about fixed rates versus variable rates. The question now is, how much does this stuff all cost?

How can we compare one offer versus another?

ALL ABOUT CLOSING COSTS

One of the harder things to understand in this whole process is the outstanding question . . . how much is the lender going to charge you?

Can you negotiate all or some of these things?

First, let me give you a resource as a benchmark. Check out www.bankrate.com/finance/mortgages/closing-costs

On this website, they show how different closing costs can be from state to state.

You see, what happens is that each state and area has different taxes and fees and ticky-tack this and that. Bankrate.com did a comparison based off a $200,000 purchase.

Yes, I know. Perhaps, in your area, that could buy you a park bench with some trees for shade and a lollipop as a thank you. However, bear with me for a second.

Some states are really low. For example, Ohio and Idaho have the lowest closing costs in the country.

In stark contrast, Hawaii and New Jersey are much more expensive for closing on your house.

Keep in mind that the more expensive your home is, the more expensive closing costs are going to be.

For example, the closing costs of a townhome will be cheaper than a single family home. Also, California, New York, and other pricier locations will bump up these costs significantly, based on the price of the home.

So, let's break this down a bit and see what closing costs are made up of . . .

Origination Fees. These are likely the MOST negotiable of your costs. Essentially, this is how the mortgage broker is getting paid. It's typically broken down into document preparation, the broker fee, an application fee, and tax service.

A good rule of thumb to use in most cases is a total of origination fees of 0.5% of your home's value. (Example: $1,500 on a $300,000 home.)

If you are paying more than 0.5%, you could shop around among more mortgage lenders, or negotiate harder with your current lender.

If you are paying less than 0.5%, you are likely getting a great deal!

The key: Don't be afraid to ask on any of these costs!

For example, you could ask the lender to waive the application fee or to lower their origination fee.

Be aware of a trick that some mortgage brokers will use: they will bump up the interest rate to make up for some of the fees that got waived. This is because they can build more into their compensation.

Third Party Fees. There can be all kinds of third party fees, and these can differ drastically from location to location. These commonly include the appraisal, attorney fees, credit report, flood certification, pest inspections, postage fees, and surveying.

Appraisal is an interesting one. You as the home buyer CANNOT directly hire the appraiser.

As a matter of fact, due to law changes over the last 10 years, your mortgage broker also CANNOT directly hire or even directly communicate with the appraiser.

They are really trying to keep a firewall between all the parties to reduce influence on the appraiser and avoid rubber-stamped deals.

Just think of *The Big Short* and how some banks worked with slimy folks to give multiple mortgages to just about everyone who would let them. Some appraisers were part of this problem, as well, in leading up to the financial crisis. They would value a home for more than it was worth.

Unfortunately, some of these, like the appraisal, attorney fees, and credit report, are not negotiable costs. Obviously, appraisers are not going to work for free; they have a business to run and people to feed.

Other Costs You Can Shop. In addition to everything mentioned so far, there are a variety of areas where you can shop & save.

For example, take title insurance. Title insurance is insurance against the risk that a title to property is not valid.

According to the Better Business Bureau, "It's estimated the average cost for the owner's title insurance policy is **$3.50** per **$1,000** and lender's title insurance is **$2.50** per **$1,000**, although the price depends on the local marketplace."

You can absolutely shop this out and try to save a few hundred bucks.

There's also the inspection. It's up to you to hire inspectors—pest inspectors and general inspectors. They are hired by you to check out the property and make sure that the property is up to your standards.

So, this is also a negotiable cost.

FINAL THOUGHTS

Buying your first home is so exciting! You can finally have the freedom to make your home your resting place.

At the same time, it can be stressful. There's so much going on in a physician's life that some of these details can get lost along the way.

Make sure to take your time and do it right. First, find the right property in the right neighborhood. Shop, shop, and shop some more.

Then, negotiate the details. Negotiate, negotiate, and negotiate some more. Don't be afraid to ask for a reasonable deal—and don't be afraid to walk away if it won't be right for you.

There are plenty of houses out there for you!

Finally, as you are ready to make an offer and it is accepted, make sure to shop out among several companies and understand what closing costs are negotiable and what closing costs are not negotiable.

The power is in your hands!

RESOURCES MENTIONED

'Why Physician Home Loans Fail' by Josh Mettle
http://amzn.to/2fzjA1X

The Physician Financial Success Podcast by Josh Mettle
http://physicianfinancialsuccess.com/

The Schiller CAPE & History of Interest Rates
http://www.econ.yale.edu/~shiller/data.htm

Average Closing Costs On Bankrate.com
www.bankrate.com/finance/mortgages/closing-costs

CHAPTER 26

3 CRITICAL PITFALLS OF PHYSICIAN MORTGAGES

By Dave Denniston, CFA

In this post-meltdown world of mortgage banking, transitioning residents and physicians face more challenges than ever.

Due to the crushing burden of an average $180,000 student debt load, and relatively lower income in residency and fellowship, physicians have a higher rate of underwriter decline than virtually any other professional. It's shocking—but it's true!

If you spend a few minutes in physician chat rooms where the topic is mortgage, you're more than likely to read nightmare after nightmare and horror story after horror story. It's so devastating to see what happens to a crushed home loan, and what closing can do to a family.

One time, out of the blue my friend Josh Mettle received a call from a young resident who was relocating to Utah. The gentleman had been two months working on a home purchase, and he had been called the day before his closing. This was a mere two days before he anticipated moving into his home with his wife and two kids.

They were literally driving across the country in a U-Haul truck and they were told that they were declined for their mortgage loan. He was wrecked. He did not know how to respond to that. He couldn't believe it. He couldn't believe that he could get so deep into a transaction, was supposed to have his keys the next day and move into his home, and at the last minute, he was declined.

What we found out was that the loan officer had failed to realize that he had about $170,000 in student loans and <u>they all showed a zero payment.</u> The loan officer made the mistake of excluding all of that debt from his debt-to-income ratio. When the loan finally made its way to an underwriter, the underwriter had not made the same mistake and eventually declined the loan.

Explore with us three pitfalls that you can address today, in order to make sure that this doesn't happen to you!

PITFALL# 1: THE YES MAN VS. THE NO MAN

In my interview with Josh Mettle, which you can find on DoctorFreedomPodcast.com, he said, *"There's a tremendous difference between a mortgage broker and the underwriter. As a matter of fact, there is tension between the two that can lead to these issues.*

We call it the biggest conflict in mortgage banking.

Quite simply, the loan officer is paid to say, "Yes." There has been never a loan office in the country who was paid on a loan after saying, "No." In order to get paid, they NEED to say, "Yes."

Think of it as the carrot and the stick.

The conflict arises because the underwriter is the gatekeeper. The underwriter is the quality control person. The underwriter is responsible, and truly, their job and their reputation as a professional underwriter is on the line with every loan that they sign off on.

That's the person who's paid to say, "No." It's not that they want to say no, but it's their job to make certain that everything fits in the box. The income, credit, and down payment have to fit. All those things have to add up and qualify within the underwriting guidelines, and it's at that point where, unfortunately, many physicians are turned down and are now scrambling to find a solution.

*Instead, REVERSE engineer the process to go through underwriting **before** you write a contract or an offer on a home. This virtually eliminates the chance that you're going to have a problem with the transaction."*

PITFALL# 2: RENTING VERSUS BUYING

Most residents and fellows transition out of their residency into fellowship in June, and are very eager to buy a new home with their increased income. On the other hand, they want to get out of the mountain of debt hanging over their heads. They grapple with the question, "Should I buy or rent?"

On one hand, interest rates are at historically low rates. Things look great. Yet, if they rent, they can save a good deal of cash flow.

We suggest that you need to be more cautious as you move and get settled into your practice. Let's consider that in most cases, you have no idea what the heck you are getting into! What if you become unsatisfied with your supervising doc or the practice, or simply don't like the city? Then, you move a few years later—you could lose a lot of money if you are buying and selling homes. In that situation, you are better off renting instead.

Also, consider what the market conditions are where you are moving to. If you're entering a market where there are multiple offers, it's heated. If you can feel there's fierce competition for properties, then that is a market where there is a lot of emotion in that market.

Whenever you see multiple offers & mortgage payments are drastically higher than rents, those are two signs to be very cautious, and if you're going to be in a home for a short period of time, you may want to lean toward renting.

Also, think how long you're going to be there. If you're only going to be there two or three years, you might want to consider renting, especially if you're not sure where you'll be after that. However, if you know you're going to be there, you should buy! Consider waiting six months, get settled in, and get familiar with your area. Just don't wait too long!

Consider this current interest rate environment, which most experts say is only going to be going up over the next 5 to 10 years.

Annually, what would be the average rate you could have gotten each year, the last 30 years, had you locked the 30-year loan? That average is 7.69 percent. In comparison, the last five years, we've been averaging slightly above 4 percent.

On a $400,000 home, that's a difference of over $12,000 in interest a year! Rising rates could easily add another $1,000/month to your mortgage payment.

PITFALL# 3: GET ENROLLED IN AN INCOME-BASED REPAYMENT PROGRAM

Thinking of rates, the one rate that we see a lot of is 6.8 percent with the student loans that a lot of physicians have. Physicians, as they are in residency and out of residency, are facing choices like IBR, PAYE, or totally deferring their loans.

For example, when you go from med school to residency, there is a lot of change going on with those student loans. They're coming out of a deferral period. You have to make a decision. Am I going to go into forbearance? Am I going to go into pay-as-you-earn? Am I going to go into income-based repayment?

Student loans are changing. Income is changing.

These danger zones are areas where loans are often declined. When you're moving in between one area and another area with this whirlwind of change, you're going to see higher rates of underwriter decline. Specifically speaking about med school and the residency, that's really where the choice is usually made whether we're going to get into pay-as-you-earn, income-based repayment, or continue to do forbearance.

That's where loan officers tend to miss this thing a lot, because those payments don't come into repayment until toward the end of the year. If you're going to finish your med school in May, start your residency in June or July—you're not going to be forced to make a decision on payments until December, usually of that year.

Many loan officers cannot wrap their heads around that change. They misjudge the qualifications, and say someone is qualified—when indeed they're not. With a physician loan, companies like Josh Mettle's look at those and we can actually qualify someone either off of no-payment if it's deferred, or in forbearance for long enough.

Alternatively, if they're going to be entering into an income-based repayment or pay-as-you-earn, the physician mortgage specialist can qualify someone off of that payment, even though they're not enrolled in it yet.

Consider that if you have totally deferred on your loan, and haven't chosen one of the income-based programs, you can still make the choice to start now!

You can possibly avoid some heartache down the road by just enrolling in these programs. If you're working for a nonprofit right now, you can start that 10-year clock on that Public Service Loan Forgiveness program, even if you may not qualify down the road.

In order to do so, you'll need to choose from among IBR, PAYE, and one of the other income-based programs. This will require you to start paying on a monthly basis, but usually the amount is fairly nominal—maybe $200 to $300 a month—it depends on the loan amount and your household income.

Consider that even if you are going not to work for a nonprofit employer, at least your loan balance isn't getting bigger and bigger and bigger with the $100, $200, or $300 you're paying toward your loans every month. Simply get her done and get enrolled in an income-based program today!

FINAL THOUGHTS

As a physician, you've made a commitment to helping others and your community.

Now make a plan to put yourself in the best possible position to buy your first home!

Make sure to remember the three keys to unlocking your first home purchase.

First, make sure you are working with an educated mortgage broker who understands the conflicts that are unique to physicians when dealing with the "Yes Man" versus the "No Man."

Second, strongly consider where you are settling down. How hot is the current market? Can you see yourself living here for the rest of your working career? Are you better off buying, or renting?

Lastly, while you are a resident or a fellow, make sure to enroll in an income-based repayment plan of some sort—IBR, PER, or other alternative—in order to prevent your debts from getting higher and higher and higher due to the interest compounding.

If, as a young physician, you focus on paying off your debts, save for a rainy day, live within your means, and put money away for retirement, you can then do the things you've long dreamed of doing, and be well down the road to financial independence.

RESOURCES MENTIONED IN THIS CHAPTER

Interview with Josh Mettle
http://www.doctorfreedompodcast.com/mettle

CHAPTER 27

3 MORE MONEY-SAVING TIPS FOR YOUNG PHYSICIANS TO UNLOCK THEIR FIRST HOME PURCHASE

By Dave Denniston, CFA

I magine . . . Picture this.

It's almost that time of the year again!

The snow is almost done falling. It's melting into big puddles for our kids to stomp in. The birds and bees are coming back out. The trees are budding. Spring is upon us.

As many residents and fellows are inking their first contract, they are inevitably dreaming of their first home. Fleeting dreams of the white picket fence, birds chirping, and finally getting out of that crowded apartment, are all dancing like sugar plum fairies in many a physician's head.

Yet many physicians struggle with the reality of actually buying a home once they have finally transitioned to practice.

They've heard the horror stories of being turned down by the bank at the last second, and don't want to deal with the stress of this transition. They have enough other headaches to worry about!

Here are two more tips to act as a preventive prescription, to block this malady before you ever transition to practice.

TIP# 1: BEWARE OF COSTLY MORTGAGE INSURANCE

One of the other things many loan officers miss for young doctors is the opportunity to avoid PMI (private mortgage insurance) or a mortgage insurance premium.

A physician can get slapped with an extra payment of PMI when they do not have 20% to put down to buy a home. The lender is required to get this insurance, because the loan is labeled as "risky" without a whole lot of equity built into it at the beginning.

This tacks on an extra payment for insurance that will likely cost at least a few hundred dollars, every single month!

First of all, let's be clear what PMI is. Conventional loans have PMI, or private mortgage insurance. Government loans like FHA have a mortgage insurance premium, but they don't call it PMI, because it goes to the federal government as insurance.

Essentially, if you're putting less than 20 percent down, you have a form of mortgage insurance, whether you get a conventional loan or one backed by FHA. That insurance is a forced cost charged to you as the borrower, but insures the bank and/or the government. This is money that you DON'T get back at the end of the loan. It's a form of a fee for being a 'riskier' mortgage.

Now, not only do you get no benefit, but you can't write it off on your taxes anymore!

In 2015, the new ruling from the IRS came down that mortgage insurance is no longer tax deductible. This extra payment of a few hundred dollars a month is really lost money.

With a physician loan, you can finance up to 100 percent, depending on what state you're in, and what price range you're in.

One way to avoid the extra cost of insurance is to pass on it, and instead pay a slightly higher interest rate. Some lenders may do this to avoid the stigma of mortgage insurance.

In this scenario, you will find that interest rates are slightly higher—you're going to pay an extra 0.25 percent to 0.90 percent higher in rate, but you're saving, compared to mortgage insurance, which is 1.35 percent.

This means that you could save a full percent annually!

In addition, remember that the interest you pay on your mortgage IS tax deductible.

Consider that most physicians pay at least 30% in federal income taxes in their current bracket. In order to truly picture the difference, you have to deduct 30% from the rates. Even the extra 0.90 percent is truly closer to 0.60 percent when you factor in the tax deductibility. This is still half of the mortgage insurance costs.

Let's say you save a whole percent a year on a $400,000 loan. That's $4,000 a year over the first 10 years. That saves you $40,000!

Lastly, consider that even if you are paying a little extra today, by two years from now, we bet that interest rates will be much higher than today.

You might be paying 0.25 percent more, but at least you've locked in a mortgage at an extremely low interest rate—4.25 percent, 4.5 percent, or even 4.75 percent.

Whereas, a year or two from now, we could be looking at 5.5 or 6 percent mortgages again, depending on what happens with interest rates.

This way, you're able to leverage a higher loan-to-value—and historically, an incredibly low cost of credit.

TIP# 2: WHAT TO LOOK FOR IN ADDITION TO THE GOOD FAITH ESTIMATE

Many young physicians aren't aware of the good faith estimate (GFE) when buying their first home.

It's a very good document to start with. Your good faith estimate is a statement by the lender that gives to you several promises—a guarantee on the rate, a guarantee on the fees, and a few other important promises.

However, there are just a lot of questions around this document. The current good faith estimate has some holes in it. It does not show your total payment with mortgage insurance, property taxes, homeowners insurance, and homeowners association dues (if applicable).

It will not show your total cash-to-close. It doesn't show anything that needs to go to an escrow account for taxes and insurance. It doesn't show any kind of credits that the seller may be giving you to cover your closing costs.

The lender says, "Here's this official government document, but it doesn't tell you what your payment is, or how much you need to bring to closing."

That's the downside of the current good faith estimate.

The upside of that document is it does quantify costs, so when the bank issues you a good faith estimate, they are saying, *It's my promise to you.*

What we suggest is this: Get a fee worksheet or a closing worksheet that's going to approximate your cash-to-close, your credit, your earnest money, and the total payment with taxes and insurance, in addition to the guarantees of a good faith estimate.

With those two things together, you can really put the picture in frame in your beautiful new home.

TIP# 3: HOW TO FIND A GOOD AGENT

My friend, Peter Kim, an amazing anesthesiologist, wrote a quick guide on physician home loans. You can download the whole Ebook at http://www.curbsiderealestate.com/.

He was kind enough to make an excerpt available for free in this chapter.

"Choosing a real estate agent can be a stressful process. According to the National Association of Realtors, there are over two million active licensed real estate professionals nationwide. There are approximately 180,000 ac-tive agents in California alone. Then how do you choose a Realtor? Around half of buyers and sellers found their agent through a referral from friends

and family. Of all the people who bought or sold homes, only ⅔ said they would use that same agent again. (Ref: National Association of Realtors)

At Curbside, we believe that choosing the right agent is vital. The right agent will be experienced, put your needs above all else, work tirelessly for you, and be an excellent communicator and educator. Sounds a bit like dating, but it's the truth - it requires the right mix of confidence, trust and chemistry.

Here are some of the questions you should be asking your agent:

- *What is the level of your experience?*
- *Are you familiar with this particular geographic area I'm interested in?*
- *Have you worked with many physicians in the past and are you familiar with physician home loans?*

Physicians in particular can be a tougher group to work with because of our schedules. The right real estate agent should be willing to work with you around your busy schedules with patience.

This same agent should also have a good understanding of physician home loans and have a good relationship with these lenders.

Competition for home-buyers is fierce and they have to be able to highlight you and your financing to the seller in order to compete against buyers who might be putting down large down payments or all cash offers."

FINAL THOUGHTS

As a physician, you've made a commitment to helping others and your community.

Now make a plan to put yourself in the best possible position to buy your first home!

Make sure to remember the two keys to unlocking your first home purchase.

First, make sure you are aware of costly insurance that could sky-rocket your payments without being tax deductible.

Last, make sure to get a good faith estimate, as well as a TOTAL fee worksheet, to ensure that you know the true costs of buying your first home, and what your monthly payment will be.

If, as a young physician, you focus on these two aspects of a physician mortgage, you will be well down the road to your beautiful new home surrounded by the proverbial white picket fence.

CHAPTER 28

AMANDA'S HOME BUYING NIGHTMARE
By Amanda Liu, MD

Dave's a smart guy. He's a nice guy. However, when it comes to purchasing a house as a resident- I (Amanda) think he's wrong.

Let me explain why.

You see I moved all the way to the US from Taiwan because my parents wanted a better life for my family. They believed in the American dream.

I moved across the world. We moved several times in the US. I moved for undergrad. I moved for medical school. I moved several times in-between then. Then, I moved for residency.

I was sooooooo sick of moving! Packing and unpacking, repacking and unpacking, packing and repacking, again, again, again, and again.

As a matter of fact, I had moved my daughter 9 times before the age of 7.

This was due to various reasons. They were mostly economic as cost of living in California was so high and that I was always trying to minimize living expenses by changing my living arrangement.

As you can imagine, when I applied to residency, I was set on minimizing the number of moves for my family.

Screw moving again!

I knew that I would be in the same city for at least 5 years for training, and most likely 6 years.

I paid a lot of attention to programs which offer internship year and residency and potential radiology fellowship after residency. I did my best to ensure that we would only move one time for the all my PGY training (internship, residency, fellowship), and maybe move for the first attending job.

Because I know that I would be in the same city for at least 5-6 years, I decided to buy rather than rent. Given the cost of owning a home is about the same as renting, I like the idea of putting away equity monthly rather than paying someone else's mortgage.

After purchasing my first home as an MS4, refinancing it as a PGY1, then attempting to refinance it as a PGY2, I'm now preparing to purchase my second home.

Having worked with so many mortgage officers from a wide spectrum of mortgage companies in the past, I can honestly say I felt lied to and talked down to 99% of the time, and hence developed a severe aversion to the mortgage industry.

I wanted to share my past woes and my recent pleasant experience with you, in hope that you will avoid the negative experiences I went through before working with the most awesome mortgage banker ever, Travis Woods from Bank of America.

FEBRUARY-MAY 2014: BBVA DOCTOR'S LOAN

I was excited to settle in and not think about moving for a while. I wanted to find a safe neighborhood with good public school and as close to work as possible. Part of me could not believe that I'm buying a home.

I was watching 3 housing markets simultaneously as I was waiting for my match results, as I know statically that most people match at their top 3 choices in the residency match process and I was confident that I was a competitive candidate.

As soon as I got my match result, which was as I expected, I narrowed in on the city I would spend at least the next 5 years in. It was nice as my 4th year medical school schedule winded down, I made a quick trip to Tucson AZ for house hunting with my dad and my daughter. We saw 25 houses in 1 day with a very efficient realtor.

My feelings of excitement didn't change, but I got more focused once I knew where I matched and I was set on moving in before the internship started.

We found the house and then had to figure out the mortgage. This was THE biggest financial decision I had to make yet. I didn't want to screw it up. I wanted to make sure that I had an expert guiding me in these decisions.

I turned to the resource that I trusted and respected and that I knew had great information- The White Coat Investor.

That was where I stumbled across BBVA.

- The mortgage banker was based in Florida, while I was in California. This was a recipe for failure. Our time zone difference made a serious lag time in our communication.

- The banker had 40 years of experience in the industry; I had 0 years, so there was a serious asymmetry in knowledge. The banker tried to be sympathetic and stoop down to my knowledge level (I was trying really hard to learn on the side, and I ran all the numbers by myself in the dark, trying to keep up). The banker got impatient at times, understandably so.

- The entire mortgage process was fear-ridden. I insisted on getting a 30-year fixed, rather than an adjustable rate mortgage (ARM).

- Results: 4.375% 30-year fixed with 11% down Doctor's Mortgage.

JANUARY-MARCH 2015: MULTIPLE MORTGAGE COMPANIES TO REFINANCE MY FIRST HOME

Here we were a year later… we hadn't moved! Amazing!

Unfortunately, with the first loan I didn't have the dollars saved yet to put down 20% equity for a conventional mortgage.

I had to get a physician's loan instead.

This meant that my interest rate was slightly higher. My doctor's loan with BBVA compass had a 30 year fixed at 4.375%. I realized that

I got a good deal on the home and that with a bit of appreciation, I was close to 80% loan to value ratio, so I looked in to refinancing. My goal was to lower my interest rate so that I could be building up equity faster on a smaller monthly mortgage.

It was accomplished, my mortgage payment decreased while my principle payment increased and all I spent was $450 on appraisal on a no-cost refinance.

- The first company I came across, Cardinal Financial, had a wonderful mortgage broker, but the numbers were not great.

- The second company I worked with had the best rates and terms, but the loan originator was plain abusive, unkind, and crude.

- I ended up signing with the second, because the numbers were the best, but developed a growing aversion to the mortgage industry.

- Results: 3.375% 7/1 ARM with 80% loan-to-value ratio (LTV), switched out of a Doctor's Loan.

FEBRUARY-MARCH 2016: MULTIPLE MORTGAGE COMPANIES TO RE-REFINANCE MY FIRST HOME

- By this point, I was practically a mortgage banker myself. I could run all the numbers on Excel and predict what the loan officer would present to me: LTV, debts, liabilities, debt-to-income ratio, APR, interest rates, and terms.

- The abuse had not stopped.

- Many of the better loan originators (i.e., responsive, transparent, efficient) didn't offer the better numbers. The originator offering the best numbers was a jerk. If brilliant doctors can't be jerks, why can mortgage bankers with good offers be an a-hole?

- Results: Elected not to refinance to a 15-year fixed loan.

WHERE I AM TODAY

- I'm selling my first home and purchasing a second one.

- The Bank of America doctor's loan mortgage officer was a breath of fresh air, from the moment we met (on the phone) to the time when my loan application was forwarded to the underwriting department.

- This banker was empathetic, transparent, efficient, something of a workaholic like me, and really good with his numbers. (I respect mortgage bankers who do math better than I do, even though to some extent I expect all of them to do math better than I do.)

- The mortgage process was not fear-anger-frustration-belittlement-ridden. This banker partners with me to make the most financially sound decision.

- Results: I'm getting a 3% 7/1 ARM Doctor's Loan with 5% down. (No private mortgage insurance, as is typical of all Doctor's Loans.)

I have to quote Amber from the popular American television medical drama, *House, MD:*

"All my life I thought I had to choose between love and . . . respect. And I chose respect. And with Wilson . . . I know what it's like to have both."

I can now say, **"All my life I thought I had to choose between the best mortgage and . . . the best mortgage banker. And I chose the best mortgage. And with this banker . . . I know what it's like to have both."**

My humor at this point is the culmination of lots of tears and rage. I hope you don't have to go through what I did to get a purchase/refinance mortgage. Set yourself up for success with a good banker. Please see my <u>recommendation page for a good banker.</u>

3 LESSONS THAT YOU CAN LEARN FROM MY HOME BUYING EXPERIENCE

LESSON# 1: DON'T BE SCARED OF VARIABLE RATES

Don't be reluctant to consider variable rate. For instance, I would have gone for 7/1 ARM rather than 30 year fixed when purchasing my 1st home.

I know at a minimum I would stay in the same city would be 5 years, but I am likely to move from this first home for my first job when I transition to practice. 7/1 ARM provides a lot of front end savings and allows me to build equity much faster and frees up some monthly cash flow too with lower mortgage payments.

Plus, if you end up wanting to keep the home beyond 7 years, you always have a chance to refinance. I did my math and realized that I could pay off the home entirely in a few months as an attending when I refinanced from 30 year fixed to 7/1 ARM in the event interest rates are high in 7 years. Sometimes, it's worth it to take a little risk with a variable rate mortgage.

LESSON# 2: DON'T BE AFRAID TO ASK QUESTIONS

Ask questions. Read. Do some research on your own. Seek advice from those who've bought a home. Even if your mortgage officer is not transparent, the only way you can improve transparency is to switch to a different loan officer or to keep asking questions until things are clear to you. Mortgages are not even 25% as complicated as medicine. If your mortgage officer can't explain the process for you to understand, it's time to switch.

LESSON# 3: DON'T BE SHY TO RENT FIRST

If I were to up and move for an attending job, I would definitely rent first. Buying a home in a different city from where you live is not conducive to making the best choice. I bought a house in Tucson when living in LA, but I got lucky that it was a decent deal. But attending homes ideally should become forever homes, more thoughts and time should be allowed to make this decision.

Plus, many people change their first attending job after three years, it's not a long enough time to be in a purchased home to not lose money on it.

QUESTIONS FOR YOU TO PONDER

Here are some questions, that I'd love to hear from you on. Please visit my page at www.drwisemoney.com and let me know your stories.

- What was your experience in purchasing in a home?
- Did you like your mortgage banker? Why or why not?
- Did you respect your banker? Did they demonstrate the same level of professionalism your patients expect from you?
- Would you recommend your mortgage banker, or do repeat business with him?

The Resident's Guide
to Contracts

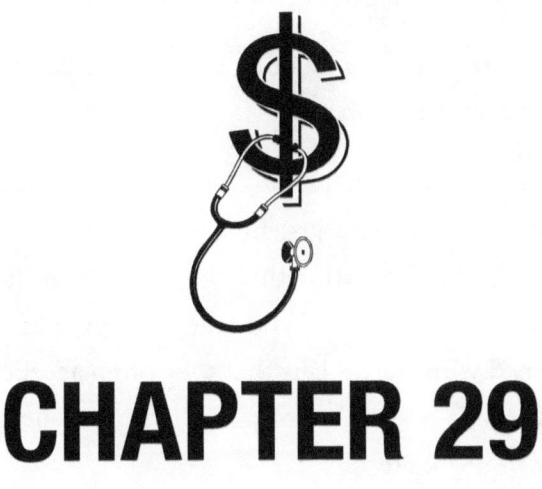

CHAPTER 29

3 REASONS WHY DOCTORS MUST UNDERSTAND THEIR CONTRACTS

You've done it! You're on the home stretch of finishing your residency.

Years of hard work are about to pay off.

You're sitting down and you receive an email in your inbox. Your eyes widen and your heart jumps into your throat.

It's an offer with a contract that requires your John Hancock!

You are about to get paid nearly $300,000 a year!!!

You dance around the room, hips move from side to side, doing a happy dance- joy overflowing as you shout, "Yes!!"

However, there's a feeling nagging at you. Shouldn't someone look over this contract and the offer?

As you think back to discussions with other physicians, the common feeling among you who just finished residency and were going through applying and negotiating through their first post-residency positions is that they are so grateful and relieved for graduating residency and overwhelmed with the process that they often accept their original physician contract as-is and do not take the time to understand it, let alone even try to negotiate.

Simply, there isn't enough time to look over this stuff!

<u>That my friends, is a mistake.</u>

We emailed a few physicians and asked them what they thought about contracts and negotiations and their number one concern. Here's what they said...

"I think too many physicians just think they get a straight salary and are good with that and they don't realize that they can structure salaries to be more tax advantaged, retirement advantaged, and advantageous to pay off taxes."

Another physician said this...

"I would say it's having very little experience with negotiation, or not realizing that there is always room to negotiate. Physicians, myself included, have accepted proposed contracts as being "standard" and good enough. Also, we're indoctrinated to believe we're not supposed to care about money, as if it's a moral failing to demand what we're worth."

Another physician had this to say...

"I wanted to write a little bit about what I found difficult with review-ing physician employment contracts. First of all, the first contract is usually the very first time that a physician has had to review and sign an extensive legal document.

This alone is overwhelming and although seeking legal guidance is completely appropriate, it is not commonly practiced (at least in what I have seen around me).

I think it is a combination of the physician feeling that they should be able to "figure this out" and also the concern that the employer would be offended if they hired legal council. I personally did not seek legal council when reviewing my contract.

I also found it hard to feel confident in what was appropriate to ne-gotiate and what wasn't. What could I ask them to change and what was considered to be poor form? This is not something that is communicated in the medical world."

We can see that there are a lot of physicians feeling this way.

They don't want to offend the potential employer. It's true. You don't want to question their integrity. You don't want to act on something that could be considered unreasonable.

This begs the questions...

<u>*Why should physicians review & negotiate a contract*</u>?

Explore and learn with us the next few chapters as you consider your future and the in-and-outs of contracts.

THE DRIVING FORCE

Unfortunately, they don't teach contracts in your training. It's hard to know what to ask, and what not to ask.

Consider this… why you shouldn't seek help from someone that does this sort of thing every single day?

Will the potential employer know you're getting help? Or maybe that you are not getting any help? It's really up to you, as a physician and how you present it.

Think about it this way…

Many employers hired a team of lawyers to draw up these things in the first place. Whose benefit do you think the lawyers were paid to keep in mind the most?

Hint: It wasn't yours, my friends!

I look at understanding your contract as a good investment of time and money. After all, we hope you are there for the rest of your career!

Without fully understanding it, or not going through it with someone who knows what they are doing, could you be leaving something on the table? It may be more of asking the 'right' questions at the 'right' time to the 'right' person – clarification is key!

What You Leave On The Table. It could be money, it could be time-off, or it could be the ability to practice medicine somewhere else close to your home in the future.

It's not that you question their integrity. However, look, you're going to spend time out of your incredibly compressed schedule & maybe your own money, good money that you worked hard for, to make sure you understand it.

Why?

Because you value them, and you value the position and you value your time. The last time that you DON'T want to do is that you don't want to go job shopping again.

In my opinion, that would be a good objective.

I see many physicians taking the attitude of 'oh well, that's just how it is.'

The conversation looks like this, "Oh, we'll just see how it goes for a couple of years, ya know. If it doesn't work out, oh well. I figure I can quit after so many days, ya know?"

How To Mentally Approach Contracts. It's all about your approach to reviewing and negotiating your contract. If you come from the mental place that you want to be there and you want to understand what is going on in the contract, you will be coming with the right spirit.

These meetings to understand the contract can clear up any misconceptions. The employer may be happily surprised.

Consider that employers may actually APPRECIATE the fact that you are so highly emotionally invested in wanting to be there. They may like the fact that you're committed & investing in the process. I've had lots of employers that I participate in the negotiation, and in fact, they thanked the employee for doing this.

To the point of the physicians we surveyed, they don't review over contracts all the time. As a matter of fact, they rarely do it, if ever.

The School of Hard Knocks. More than anything, I don't want to you to learn these lessons the hard way- to avoid enrollment in Hard Knock Life Experiencies University.

If we all lay out our cards on the table & all the expectations are put out in front, this can help everyone involved.

That means the employer can help prevent physicians, who sign a poorly written and explained contract, to not be ticked off when two years later they realize what a bad deal they got.

The employer can make sure to improve retention rates and you'll stick around longer and longer.

Let me boil it down to three specific reasons why residents should understand and then negotiate their contracts.

REASON# 1: CLARIFICATION

Make sure you have clarified how the contract works and what it says (and does not say!).

Be Crystal Clear. Contracts are simply to set expectations. How much they'll pay you and what do you need to do for the money. You need to understand how you are going to get paid. For example, how would you earn your productivity bonus? When you could you earn partnership? How do the clauses and claw-back portions of the contract work?

Additionally, what if you want to start your family? How does maternity & paternity leave play a role?

We want to make sure we ask for all the things that you want and we've addressed numerous scenarios to understand the consequences of what is clearly stated (and not stated in the contract.

We'll address specific components of the contract later.

REASON# 2: TRANSPARENCY

Since you are intending to be working with this employer for a long time, we want to have an open and honest relationship.

In being willing to review over the contract, consider the following questions:

> How willing or how open are they on discussing provisions that are important to you?
> How are they going to respond if they are in a dispute?
> How are they going to respond if there are some things that may need to be discussed?
> How, if you have a concern with your practice, or your patients, or your nurse, or your staff, or your, ... how are they going to handle that discussion?

As you engage in reviewing over the contract with them you can get a vibe from them on how they respond with what you ask them.

If we fast-forward these little questions to ask while you're not yet employed, it could be a good indicator of how willing they will talk to you when there is a problem.

Sometimes you can learn a little or a lot just by asking the question. As a matter of fact, you may not care about the answer. The reason is that you want to see how they are going to respond to it, when you ask them.

At the end of the day, these issues need to be discussed, so you have good expectations on everything, around the position, and the relationship.

REASON# 3: BEING ABLE TO SHARE CONCERNS

Above all else, you want to know that you are able to share concerns with your potential employer. You want to see all that you can discuss with them.

- ➢ What is being considered taboo?
- ➢ What are they not willing to discuss?
- ➢ What are they uncomfortable with?
- ➢ Are vital issues that are important to you being left out?

You should feel confident and comfortable going to your employer and talking about concerns that you have.

That should be true whether or not it's a contractual concern.

What if you are working there now and it's a concern about another employee? Maybe instead, it's a concern with your ability or your production.

Having that ability to go to them and talk, is important as well.

FINAL THOUGHTS

You've been working towards this destination, your whole journey in pursuing medicine.

Perhaps, this was a dream of yours from when you were a child, your under-grad, medical school, and residency.

Enjoy the Process. This is your time to be a free agent. It's good to be grateful for being wanted.

The reality is that there's a physician shortage out there in most areas in the country and while each situation is different, employers use contracts to bind physicians to their position and control the relationship.

You are valuable and you are in a position where you have the ability to negotiate. You're not just another corporate bee.

All of that being said, let's tell the truth here- with some contracts in some areas- they just don't change them. It ain't going to happen. That's what you get sometimes.

However, it's not always about having something change in the contract. It's about clarifying, and discussing. It's about being confident you can do that. And you 'don't know what you don't know' having never done this before (or having done it once or three times).

At the end of the day, we all have the same goals. You would like to work at a given employer in a given area, and they would like to have you. The employer has patients that need care. You can help provide that care. It's just how do we get understand the contact and have everyone happy with the same results. This prevents surprises and 'well I didn't understand that' type situations.

You can do this! Take the next step and read the next few chapters to understand how contracts work, what you need to pay attention to, the questions you should ask, and the pitfalls that you can avoid.

CHAPTER 30

EMPLOYED VS. PRIVATE PRACTICE VS. INDEPENDENT CONTRACTOR

Contracts are nothing but expectations – of you and of the employer/contracting company. What are they wanting from you and what will they give you? It is easy to say 'what are they going to pay you and what do you have to do for the money?'

These details need to be in the agreement – the contract has many more important and relevant pieces but no matter what type of employment you are seeking, these core things need to be included. To that end, there are very different types of employment or work for physician out there.

You can join a large regional (or even national) employer, start your academic career in a University setting, or go the private practice route.

Each has it's benefits and risks – but one thing remains the same…the contract is there to set expectations.

It's important to know the differences between them and how to view the situations. You can/should ask for different things in each situation. This chapter will discuss each one and provide some tips and pointers.

EMPLOYED

Last year over 65% of Physicians out of training signed into an 'Employed' opportunity. No partnership offered – they'll be an employee for the foreseeable future.

Often times these opportunities are presented to candidate in a longer, more 'standard' contract. It may be 12-40 pages long (we've seen 76!) with addendums and exhibits. Often the contract itself is somewhat standard and the exhibits are modified depending on your situation and position.

Compensation and duties may be detailed for you differently than a colleague (think Interventional Cardiology or Trauma Surgery vs Shift-base Hospitalist). Often times physicians wonder if these contracts are standard and non-negotiable.

Do they even need to be reviewed? While they may be a little less flexible than a smaller private practice (each situation is very different) there are very important points you need to know to work through these agreements.

Large employers may have multiple sites to practice at. It is very important the agreement is specific with your location. This is often a detail that is left open. Contracts often say you'll work where they choose.

While we always consider intent (we don't think they would verbally tell you one thing and do another) this is not always the case depending on their business needs. Their business and patients come first – not your happiness or job satisfaction.

It is important the agreement states a specific address and gives you some control over any changes. This may even have an effect on the non-compete section if one is included.

To that end, non-competes are often in these agreements. It is important to understand it but also know if it is standard for all. We encourage our clients to ask many questions around these restrictions while discussing/negotiating their agreement.

Can it be modified? Is it standard? Does everyone have the same non-compete? Is it from my primary location only or anywhere I work? How does it terminate or become enforced? Have you had anyone violate it in the past? What did you do? Have you allowed a physician to breach it and not enforced it? What was the situation?

Of course, lower time frames and geographic areas are preferred. Restrictions from your primary location we feel is a must. It should also appropriately dissolve upon termination for any reason out side your control (typically a no-cause by the employer, for cause by you, the Physician, or a non-renewal of the term). We feel these small and very reasonable requests help to balance the risk you have in these very important restrictions.

Of course, know and understand the compensation structure. While these may be more static than other situations, they may be very flexible depending on your situation (the 20th Hospitalist to the service line or the only Reconstructive Urology Surgeon in a large catchment area).

Know what you are worth and be able to request appropriate compensation for the work you'll be doing. Be creative with a balance of base salary, bonus, signing and relocation amounts as well as student loans (student loans more common in these employed situations).

We typically provide our clients with multiple creative options for negotiation depending on how flexible the employer is. We feel at the end of the day it is just a good practice to have many questions around this to set the appropriate expectations. Get creative!

So while these employed situations ARE negotiable and often times very much so, there are more standard parts and language and their corporate attorneys may be less likely to modify the agreement. Various 'standard' parts often include the termination provisions, benefits offered (including malpractice limits and type of policy), and overall structure/language. Items that are often negotiable are compensation, non-competes and 'tail' insurance, and job duties/expectations (including location and schedule). We feel there are no 'standard' or 'boiler plate' agreements that don't warrant many many questions and most likely, slight modifications or negotiations.

PRIVATE PRACTICE

While the large employer contracts are somewhat flexible, often times private practice agreements are open to many different modifications. They may be shorter than the employed situation and often range from 5-20 pages (we've seen 1 and it was NOT easy and 'simple' as you may think!).

Shorter agreements are NOT typically better as many physician think. They often lack the detail necessary to set all the expectations from both parties (again, what a contract is for).

It is usually easier without the standard structure of hundreds of physicians signing the 'same' agreement and typically the practice owner (not a team of corporate attorneys) are able to make changes or agree to them verbally for their small practice attorney to make.

These agreements often just need more work and negotiations than the employed contract anyway. While many of the same points from the employed section apply (location, schedule, compensation) the negotiation process is much different.

You are often speaking directly with your future partner or colleague, not a recruiter or Vice President as in the employed situation. If there is potential for partnership in the position, clarifying the 'what' and 'when' is one of the most important points. While starting or guaranteed compensation of private practice typically lags that of a large employed situation, the future of being your own boss, owning your own practice/business, and making more than the employed physician is often very alluring to younger physicians.

The discussion around Partnership should be very purposeful and calculated. Have your questions before you begin the conversation. When is partnership offered? Have you ever not offered it to anyone in the past? Has the offer changed over time? How does my compensation change? What does it cost? What do I get for the buy in? How would anyone transition out once a partner?

These are just a fraction of the important questions you should have top of mind when evaluating these opportunities and agreements. The details may not be included in the agreement, and that is normal.

Owners typically do not want to make any guarantees before they know you and your skills, this is why it is very important to understand the situation and ask the appropriate questions during the interview/contract negotiation process.

Understanding the intent of the owners is also important. While you may thinking you are signing and in 2-3 years you'll be a partner much can change. There is great consolidation in the medical space currently.

Some groups are being bought by private equity and others integrating into a hospital system for different reasons – it varies greatly by specialty and location. What happens in the event of a buy-out? Has the hospital been approached in the past? What would they say? How would I be affected?

There are many important questions to ask. Again there is little a contract can do to protect you in these situations. You can ask for notice should they begin talks, or a provision that the agreement is/will be assignable in the event of a merger or acquisitions. These are tricky situations that should have the eye of a professional on your side to help.

Compensation usually tends to be more negotiable. Know how they are paying you (salary, bonus, etc) and if your collections and any threshold amounts for a bonus is reasonable. These vary a tremendous amount for many reasons – location – specialty – size of the private practice – etc.

If the compensation structure includes expenses you need specifics. These agreements often lack detail on what you'll be 'charged back' for expenses if you are paid on collections – expenses.

These could vary greatly and often times you may run a negative balance against a 'draw' that would need to be paid back should the agreement end. You need specifics in the agreement on all expenses you are responsible for (benefits? CME? Overhead? Direct and indirect expenses? Etc). Know average collection times, know how you are paid any bonus upon termination. It is important you understand these difference before asking to negotiate anything specific.

Many of these specifics should be included in the agreement and not just verbal promises. Should there be a termination event or dispute around how a bonus or expense is calculated, not having them in the agreement could be costly.

Private practices may have a much different benefit structure than larger employers. Know what costs and fees you are responsible for. Life/Disability insurance and the medical insurance can vary greatly. Understand the details, how changes can be made, and when the particular benefit 'kicks in'.

Malpractice is more likely a 'claims' policy and a 'tail' will need to be purchased should a termination occur. Most often in a private practice this expense is passed on to the physician. This is a great negotiation point. Know the costs and structure and provide some good reasons why they should cover it (especially if the agreement ends for any reason out side your control). Should they not offer partnership in a defined period of time, we prefer if they cover it. Again, we are just balancing risk should you take less money for a potential future as Partner and that future not be offered or presented.

One final piece on negotiating private practice agreements. Having the right questions at the right time is important. They are much more likely to become upset and withdraw an offer if they feel the question are inappropriate or you are seeking things that are unreasonable. An employer that withdraws an offer simply because you are asking purposeful questions may mean additional red-flags depending on the situation

We often get the 'I never had a signing bonus so why should we give one?' even if they signed 20 years ago and times/markets have changed. The 'trust me, it doesn't need to be in the agreement' can be more tricky to overcome during these negotiations. Again, very intricate situations that should be taken in a purposeful manner to ensure it gets done right, but with respect to all parties. Anticipate their answers and objections and plan the discussion accordingly.

ACADEMICS

Typically Academic agreements are not as flexible as a traditional employed or private practice agreement. They are often written in a 2-3 page 'letter' and not a formal contract. They tend to reference many policies and provide no detail – even on important items like termination and bonus structures.

While they may be less likely to modify or change these agreements, it is just as important (maybe more) to have great questions to ask as you discuss and negotiate these agreements. Know all the policies that are referenced. Understand their termination process if it is not included.

Know how they set the salary and if it is negotiable. Tenure? Benefits? Funding source? While not as flexible or negotiable there are many items to discuss.

Again it is all dependent on your situation. The 12th pediatrician or the first Kidney Transplant Surgeon to start the program? You may have different needs and negotiation points/capabilities even in these positions.

Schedule and location still remain important. You should have these defined in the letter they provide. Typically they offer an occur-anced malpractice policy so no 'tail' is provided but you should know and understand all these details that are usually not included in the academic agreement.

Tip on Academics – many Academic services are re-doing their physician services under a separate entity. There are many reasons for this but we are finding many physicians that are signing 'academic' positions are more likely going to receive the letter typical of these jobs but also a formal 10-20 page 'contract' for the physician group. These should be treated like an employed situation and discussed/negotiated as such. You are more of an employed physician vs a 50% clinician and 50% teacher/researcher as a true academic position would be.

TIP! Academic agreements are still agreements – you need to know your requirements and flush through any potential surprises that may come up. Usually the lack of detail is not as big of a deal as in a private practice agreement. However, knowing and balancing the risk, even if typically less in these agreements than others, is good due diligence on your part.

1099 INDEPENDENT CONTRACTOR

There has been a big shift and trend to more independent contract workers in the physician market. While still much more common for certain specialties (Emergency medicine) we are seeing a trend towards more 1099 income and less W2 for physicians. There are many benefits but risks as well.

These agreements are also most often flexible and negotiable. Often times the most important item in this situation is not a contract negotiation point – it is knowing your tax situation and what you can/cannot do as well as what the entity offering the agreement can/cannot/will do.

Find a good CPA that understands your area and grab some good advice on how to set up your taxes. You should ask many questions of the company on how they pay and what your obligations are from benefits to expenses. You may have certain things provided (malpractice insurance, budgets for costs and fees, etc) or nothing at all.

You should ask about the contracts they have and when they expire.

If your relationship is more short term or likely to change, understand your outs and how the relationship can terminate (including the company losing their contract with the facility). Understand the history and other physicians that may be contracted there.

The compensation is often negotiable for 1099 agreements.

You should know how you are paid on termination for any reason and make sure you understand and negotiate any restrictions or obligations post termination to minimize your risk. They are not employing you or looking long-term into partnership. These are reasons you should view the near-term risk mitigation of these agreements to be one of the most important items on your negotiation list.

So there you have it – a tiny – small – fraction on what is important with these deals (and we took up 5 pages on just some of the important things!). While each contract is different and unique there is always one thing in the background – your situation and story. Some want to stick in the sitation for a year and move – others are there long term. Others are joining an existing practice or taking one over yet others are building a new one from scratch. Either way, the contracting process is very important. After all you've spend hundreds of thousands of dollars to train yourself over years and years, dedicating yourself to your craft. There is just too much on the table to have things missed or forgotten during this important time.

CHAPTER 31

10 COMMANDMENTS OF PHYSICIAN EMPLOYMENT CONTRACTS

There's nothing that jumps my heart and my adrenaline like hearing a police siren.

Can you remember the last time you heard a siren?

The wailing moan of the siren piercing you ears as it draws closer and closer.

Can you hear that sharp wail as you close your eyes for a moment?

Unfortunately, there are folks who break the law & we're so grateful to the law enforcement community who protects us.

As you go down the path of understanding and maybe even negotiating your contracts, I've boiled down 10 rules of the road, commandments if you will, to keep you on the straight and narrow and avoid horrible accidents.

We will reveal each one of them to you and then briefly describe each of them to you.

THE 10 COMMANDMENTS OF EMPLOYMENT CONTRACTS:

1. Thou shall understand how to get out.
2. Thou shall understand all forms of compensation. Salary, signing/commencement bonuses, benefits, production/quality incentives/awards, retirement,
3. Thou shall know the location and schedule (including call) and if it can change.
4. Thou shall understand all the 'nons' – non-compete – non-solicitation – non-interference – non-disparagement
5. Thou shall know any pre-defined damages/penalties (i.e. liquidated damages)
6. Thou shall understand the medical malpractice policy (limits, type, tail, changes)
7. Thou shall know their market value.
8. Thou shall understand all additional agreements/contracts affiliated with the practice or position (hospital agreements, PSA's, income guarantees)
9. Thou shall know their worst case, best case, and BATNA before signing.
10. **Thou shall review the previous 9 commandments!**

THE FIRST COMMANDMENT: THOU SHALL UNDERSTAND HOW TO GET OUT

Most doctors' intentions are going into a good position, honestly for the long term.

However… what if the position doesn't work out?

Consider these questions:

➤ Do you have a "No Clause?" termination?
➤ How can you be fired? Will there be any notice given?
➤ How you can be terminated or suspended?
➤ How and when does the contract end?
➤ Is there an automatic renewal? Is the term length acceptable?
➤ Do you have to renew your contract every single year?

Those are all good questions you would want to ask to know how to get out of a specific contract.

While most physicians start employment with long term intentions the data shows otherwise. Multiple studies have shown over 50% of physician's first job out of training will be only for the first term only. The likelihood that you transition to a new practice is high. Dissatisfaction among physicians is also at an all time high (among certain specialties). These factors show how important it is that you understand how you can leave your current practice. Most agreements have a no-cause termination provision by either party. These can typically be 60-120 days but may be shorter or longer. Think of this question:

How long will it take me to transition into a new practice?

The items that need to happen are many and include wrapping up your current practice, finding a new opportunity, applying, conducting a site/community visit (maybe two), obtaining a contract, having it reviewed and negotiated, then becoming credentialed, licensed, payor approved, and potentially relocating and a new state license. This entire process can be well over 120 days. It's important to understand any gaps in income you may have over the course of this transition. This is the reason we prefer to have 90-120 days as a minimum for the length of time you'll be paid on this type of termination.

These agreements can also typically be terminated (maybe immediately without notice) if you cannot practice due to a disability, loss of license, restrictions on practicing or billing, or a host of other items. It's important to understand any of these immediate termination provisions and if any are vague or have the potential to be misunderstood or interpreted. *How does the employer define disability? What are all the 'policies' that are referenced? Has there ever been disputes?*

Having a specific reason to terminate with notice can also be important. The contract sets expectations, so not meeting those expectations (your work schedule, their payments to you, etc) violates the agreement. If a party is not keeping up with their end of the agreement, the other party can request they do within a defined period of time, typically 10-30 days, or they may unilaterally end the agreement with cause.

THE SECOND COMMANDMENT: THOU SHALL UNDERSTAND ALL FORMS OF COMPENSATION

Physician compensation has changed dramatically over the years and will continue to do so at an ever increasing pace. While most physicians have hundreds of thousands of dollars in debt when they start their attending career, understanding how they are paid is vital. Upfront cash compensation is common in most situations and can be in the form of signing bonuses or commencement payments. There are often details on how and when these are paid as well as repayment provisions in the contract. What is typically not included is how and when they are taxed. If these amounts are structured as 'loans' they may be forgiven over time, which may incur a tax-liability at that time.

Most physicians have a base salary and these are usually straight forward. Bonus provisions can be very intricate and are dramatically changing in todays current healthcare landscape.

'Incentive Compensation' can mean many things. It can be on many forms of production or quality. There can be very specific metrics or much vagueness. They can be paid often or rarely. They may even be up to the 'discretion' of the employer! It is vital to know all the details around any bonus pay – what the metrics are – are they the same for all? When is it paid and are there any policies around these payments? Understand the expectations and if the goals they have set for you to receive the incentive are even reasonable or able to accomplish. We could write an entire book on the do's and don't's of what to look out for in your compensation structure. For this commandment we can only encourage you to ask as many questions as you can – ask us if you like – but do not be one of the many physicians out there that doest' understand how they are paid or what happens should they cease to work with/for their current company. Knowing what happens on ter-mination/transition can be almost as important as how you are paid while working under it.

You could be paid per shift or per hour. Know what the expecta-tions are for the shift work and have any guaranteed hours delineated in the agreement. This can be important for when you are trying to consolidate student loans or take out a mortgage as these lenders need to know how much you are going to make. $200 per hour? Well it matters if you are guaranteed no hours or 120 hours per month! Have these details included in the agreement and it will make everyone's life just a bit easier.

The fact is compensation is changing and there are entire books on this subject alone (we may write one!) At the end of the day you need to know how you will be paid, when the payments will be made and how you are paid on any termination. Understand the expectations and structure. Understand how others are paid. Understand how and when things could completely change and you should avoid any surprises in the future regarding something as important as your income!

THE THIRD COMMANDMENT: THOU SHALL KNOW THE LOCATION & SCHEDULE INCLUDING CALLING

Sounds pretty easy doesn't it – where am I going to work and when shall I show up? Unfortunately these details are rarely provide in great detail in physician agreements.

How many locations are there? Is it multiple? Is it just one? Could it change? While we understand these are discussed and typically agreed to during the site visit, business and patient needs can change at anytime. The Administration can change anytime – and if your contract is not specific with these details your responsibility, requirements, and job satisfaction could change anytime as well.

We've said it before – often times what isn't in a contract can be as important as what is. As medicine and practices consolidate, employers seek to increase the scale and scope of their services and locations, the expectations around where and when a physician works can be very important details. Contracts should state the specific location and schedule of the physician. Having a 'minimum' number of weekly hours or undefined hours or times (days/nights) shifts can be risky to a physician. While the practice needs can change the employer should be required to obtain approval from the physician for any specific changes to the schedule or location. Call should also be defined as best as it can. Capping the amount of call a physician can take in a month can prevent physician burnout, which is good for both the employer and employee. While all parties have the best intentions when going into these agreements, this is one area that is consistently vague when it should be clear. A physician takes a job based on pay, the schedule he or she will work to earn the compensation, and the location he or she will live. Having the location and schedule defined is just as important of an expectation as the compensation (which is of course always defined).I would assume that with a lot of happy physicians out there that are working under great contracts with a great employer, and everything is good.

THE FOURTH COMMANDMENT: THOU SHALL UNDERSTAND ALL THE NON-COMPETE, NON-SOLICITATION, NON-INTERFERENCE, AND NON-DISPARAGEMENT.

Knowing what you can and cannot do both during and after your relationship ends with the employer/contractor are of course great importance! Most contracts contain post- termination restrictions. Sometimes they are as simple as stating you cannot contact or solicit the hospital you are currently providing services at. Other situations restrict physicians from contacting patients, employees, vendors, and then from working in large radiuses from multiple locations. These provisions should be a little flexible with how the agreement terminates – anything outside your control and they should be forgiven. This just balances risk appropriately we feel. Think about it, the empoyer hires you, and decides they don't need or like you so they terminate you w ithout causeThey let you go without cause? The group is bought by the local hospital? You should not be restricted if you didn't play a role or are the reason in the reason you are no longer employed there. Not knowing these provisions can be potentially very costly and career limiting. Many contracts contain damages you are agreeing to pay them by signing the agreement. These damages may be as high as your last years W2 earnings. Be aware of all the non's and restrictions in your agreement at all times and how they may change over time.

THE FIFTH COMMANDMENT: THOU SHALL KNOW ANY PRE-DEFINED DAMAGES & PENALTIES (I.E. LIQUIDATED DAMANGES)

We can make this simple - Liquidated damages are almost like saying instead of going to court to fight it out in front of a judge or jury, to figure out right and wrong.

You'll just agree up front in writing by signing this contract that you, the physician, is in the wrong.

As you look to minimize risk in your particular contract you need to be sure there are no pre-defined damages. These are often called 'liquidated damages'. They represent an amount defined and agreed to at the time of signing should something specific happen. The employer does not need to prove the damages, and any actual damages to the employee may be significantly less than the predetermined damages. If there are certain breaches or violations our court system allows a process to determine what (if any) damages have occurred. What you are doing by signing an agreement like this is bypassing this process and agreeing that the penalties should be a certain amount. While very common in certain areas and situations, these are less than idea for an employee.

Therefore, if you do breach the non-compete, or any provision with predefined damages, they can send you a bill for $200,000 or $400,000 or even $500,000 if it is in the contract or whatever the pre-defined damages is ours.

Will they actually enforce this provision? It's really hard to say.

Minimizing the risk in your overall agreement includes having a well thought out discussion around these provision and requesting removal or modification.

THE SIXTH COMMANDMENT: THOU SHALL UNDERSTAND THE MEDICAL MALPRACTICE POLICY (LIMITS, TYPE, TAIL, CHANGES)

Medical Malpractice Insurance is something nearly all physicians generally need (certain government employees may be exempt from liability) and should be defined in all contracts. It is typically an employer paid expense but often with many moving pieces. The limits may be high or low – the coverage occuranced or claims made. While there are many details on the different policies the basic points are as such

Claims policies cover you for claims MADE while the policy is in effect.

Occuranced policies cover you for when the claim OCCURRED.

So, if you have a claims policy and you leave the employer, the policy will be canceled (just like if you sell a car, you stop the insurance policy). So claims that are filed after the cancellation of the policy will not be covered. This is what a 'tail' or 'extended reporting endorsement' is for. Depending on your policy you may or may need one. It could run you $3,000 or $180,000+ depending on your area and specially.

Even if you are covered you should know how the policy could change. Is malpractice listed as a benefit? Just like the employer may change their health insurance coverage year to year they may (depending on the contract language) change the policy and structure. Know this very important 'loop hole' in some agreements.

Maybe you are a resident who is reading this book and you are open to going to many different places. Simply, you just want to find where ever you can find the best deal.

Specific limits and needs vary by state and specialty. Know your local market and what the other physicians in the practice have before you sign the agreement.

THE SEVENTH COMMANDMENT: THOU SHALL KNOW THY MARKET VALUE

What are you worth? Think about it. You've invested the time and money; this is a very important question to have a very specific answer to. Most physicians know what they WANT to make but not what the market will provide. Just like housing prices there is great variability on how much physicians make – of course both per specialty and per location. We see Neurosurgeons and Surgeons making well into seven figures in some parts of the country and lower $300,000 in others. Same or similar job, much different pay. We've done Pediatrics for $275,000 per year and $145,000 per year, 110 miles apart! Many factors come into play and it is important to know and understand as many as you can.

What should your wRVU rate be? What is the average overhead or collection rates in this part of the county? What should a base salary or guarantee look like? There are many misleading websites and surveys out on the internet. Many have large selection biases or small numbers of participants (especially in some specialties). If you don't know how much you are worth, how can you evaluate one of the most important parts of the agreement?

As physician compensation changes so does this very important dynamic. While you want to make sure you are maximizing your income and salary during the negotiation phase, you want to know that what you are asking for is reasonable and within market rates. Discrepancies do exist – American vs foreign trained physicians, male and female, old and young. You've worked to hard to get to where you are to not only know your current and future value but be able to have a reasonable conversation with your potential future employer around how they set the structure and what is able to discuss/modify/negotiate.

We've talked a lot about market value and geography earlier in the text. However, let me emphasize that you need to know what type of position you're looking for.

For example, let's take a Pulmonary Critical Care position. Let's say that there were two opportunities. The salary difference was two-fold. One is $160,000.00 a year and the other $320,000.00 a year.

Big difference! HUGE difference in compensation!

The lower paying position was an academic position where the appeal was the quality of life was more balanced and this physician could have the prestige around an academic career. There's just a lot of great things about that job that don't revolve around money.

You get to do research, you get to teach, and you get to educate. Unfortunately, in this example, you don't get paid as much.

In comparison, the higher earning position is in a private practice world. However, you'll also work quite a bit more. You'll have a much different level of stress and a much different level of job satisfaction.

The schedule and flexibility as well as the prestige that comes with affiliating in an Academic Institution is a wonderful thing for many – but not for others. Know what you are looking for and create the job search appropriately.

There's a big difference on wanting to be a corporate employee where you show-up and there's no decisions made on a business. There's no profit or loss to look at. There's no contracts to negotiate or payers, or no nurses to hire or business associates to plan for. You just show-up at work and go to work.

Do you want to be an academic and publish lots of journals and do research, and try to advance certain therapies there? Or do you want to be in a private practice and walk away with equity and more compensation?

All of these are very important considerations on compensation.

Again, I can't stress enough the importance of considering where you are practicing.

If I'm a resident and I have $300,000.00 in debt, my decision on where to settle in and practice would likely be much different than if I am a resident and I have NO debt.

If I'm a resident with loads of debt, I would be looking at some of these other places to live like Alaska or North Dakota or someplace where I could be paid significantly more to pay off my debts quicker.

Certainly, family and a strong support system play a very important role in this decision. However, keep in mind the financial implications of making this location decision.

The compensation of where you practice could literally be in the six-figure range!

You could easily be compensated $100,000.00 more by living in 'less sexy' areas that are away from the coasts.

Keep in mind that in addition to the position of compensation and then obviously cost of living, something completely different. As far as the, if you want to call it, "The quality of living life." As far as how much further those dollars go, can be very, very dramatic, and different as well.

THE EIGHTH COMMANDMENT: THOU SHALL UNDERSTAND ALL ADDITIONAL AGREEMENTS & CONTRACTS AFFILIATED WITH THE PRACTICE OR POSITION (LIKE HOSPITAL AGREEMENTS, PSAS, INCOME GUARANTEES, AND SO ON)

A lot of times you'll see contracts that will reference other agreements. Obviously, it's important if the contract references other agreements that you are agreeing that we know what those things say.

I ask sometimes physicians their salary, or their income that's verified by the hospital, or from a private group.

It's very important to know how that is set-up, to make sure that there isn't any financial obligations on the physician that they're aware of such things.

THE NINTH COMMANDMENT: THOU SHALL KNOW THE WORST CASE, BASE CASE, AND BATNA BEFORE CLAIM

Anybody who's negotiated, or has read lots of books on negotiation.

What is a BATNA?

Your BATNA is your – Best Alternative To a Negotiation Agreement.

If I'm a physician, and I have five offers, and I don't care where I go, I just want the most money.

It's easy for me to say, look either you pay me this, or I'm out! I'm going to walk away.

However, if I'm a physician, and my only goal in my entire life is to go back and work in a certain town, my negotiation cap may not be as great.

My best alternative to a negotiated agreement means asking the question what's your threshold for walking away from the deal?

What's your best alternative if I don't have a job in my ideal location?

In this scenario where you want to work in a very certain location, you probably don't have much threshold for saying, look, pay me this, or I'm not going to sign.

If you've got five other offers and you don't care where you live, you have room to make more negotiation. Simply, there's low supply and high demand for your services.

Who Wins the Negotiation. To be blunt, the person who cares the least generally wins.

If you don't care? I'd say the deal works out. You're probably more likely to get what you want. If not, who cares? You walk away and find the deal that works for you.

The SUV Example. Consider, this example… let's say that you want to buy a black SUV and you don't really care what kind of SUV it is.

You can call around to all the dealers and say, do you have any black SUV's?

If you don't care about the type of SUV or the year or what's inside it, you have POWER in negotiation.

You can walk away from the deal no matter the make or model. You just want the best price with a certain amount of miles and you are open to seeing many different types o vehicles.

In contrast...

Let's say that you HAVE to have a cream colored 2014 Cadillac Escalade with seat warmers and a DVD player and 6 coffee holders, your desire to be so specific REDUCES your negotiation power.

FINAL THOUGHTS

My friends, remember that your future is in YOUR hands.

You now have a working understanding of the 10 Commandments of Physician Employment Contracts.

Review over them and make sure that you come to the negotiating table with as MUCH power as you can muster.

Now that you have a really good understanding of how contracts work, the next step is to understand how contracts and backfire and to learn from the mistakes of other physicians.

CHAPTER 32

CONTRACT HORROR STORIES
By Dave Denniston, CFA

Here you are - transitioned to practice and you can't believe it! You whistle a happy, tuneless ditty to yourself in content. The paychecks are a tidy sum of dough. Those dollars are flowing in.

You're working towards buying your first home and your student loans are finally going the right direction- DOWN- rather than up.

However, you have a nagging suspicion that something is amiss.

As you take the time to think about your job... you wonder, can I really keep this up forever? I'm working insane hours and I'm not sure that I can keep going at this nasty pace.

You take the time to review over that contract you quickly signed and you find out that you are screwed...

Your heart thumps so loudly you swear it is beating outside your chest and a cold sweat forms on your brow.

"Oh boy, what have I gotten myself into???" you wonder aloud.

This next chapter is dedicated to the countless physicians who have been confused and made mistakes with their contracts.

Learn from their mistakes. I find that successful people learn from their own mistakes, but SUPER successful people learn from other's mistakes. Mistakes when it comes to a physician's career, earnings, etc can be very costly. It can be costly in terms of time and in terms of money. Often times, both.

These are stories that have been told to us – or we have helped those in them to get out and find a better path. All the information is true and accurate, but any confidential information has been omitted for obvious reasons.

Read, learn, and grow...

STORY 1 - THE ENT PRIVATE PRACTICE NIGHTMARE

There are two types of non-academic physicians these days. Those that are interested in private practice and partnership tracks and those that want to start and remain an employee.

With the number of privately owned physician groups shrinking dramatically over the last few years, those physicians looking for these opportunities must complete extra due diligence when evaluating offers. They can be extremely rewarding from many angles – but also carry more risk than a typical employed situation.

We often review employee – to – partnership agreements and then the actual partnership paperwork years later. There are many items to consider before beginning what may be a two to five year journey on the way to partnership. You may consider this 'lost time' if things do not work out as you expect and you need to start the process over again. This can not only be frustrating for a newly minted physician, but financially costly as well.

First, having a discussion with the physician owner/partner up front on what you are looking for and what he or she is offering is very important. There are many questions that asked up front, can save much headache and lost time in the future. Just a few of these question could be as follows

- How are they viewing the process and future relationship?

- Are there certain metrics that need to be met in order to be offered the partnership?

- What general terms are going to be offered?

- Have you ever hired a physician and not offered them partnership in the past?

- Is the potential partnership opportunity going to be equal or will you retain a majority share?

- Are there other ancillary revenue streams such as real estate or Ambulatory Surgery Center revenues I'll be able to have access to?

- What would happen if the business suffers a major even between now and then? (their disability/death, a lawsuit, a buy-out, etc)

- Has anyone approached you with interest to buy the practice? If so what was the discussion?

- Is that something you'd be interested in the future? What would my role be?

Now many private practices may know all the answers – some may know none – it all depends if you are the 15th partner or the first. Either way, it is important to ask the right questions to the right person at the right time. We'll explore these details as we discuss a recent interaction with a practice owner/physician that was looking on bringing in his first employee/partner.

There was a recent employment contract we reviewed for a private practice Otolaryngology position in a major metropolitan area. It was a solo practitioner that was busy enough to hire another physician and had the hopes of adding a future 'partner' to their successful company.

Many Locations. The practice had 3 current locations and multiple referral sources as well as a partial ownership in a local ASC. It was a multi-million dollar business as the entrepreneurial owner had been successful over the past few years with many late hours of business development and planning in addition to being the solo physician in the group.

Our physician client was a highly trained surgical specialist that would have been offering a service not currently offered by the practice and the owner of the practice was very interested in our physician client. After a review of the proposed contract, the physician elected to have us do the discussions directly with the group.

Many Purposes. The goal of contract negotiation/discussion can be multi-purposed. It is not as simple to state the negotiations are to get the 'best deal' possible whether that mean financial or term or otherwise.

While we (and you should too) look at the negotiation process to include obtaining the best financial terms, there can be much more to the process. Making sure you understand your 'out' should things not work out and having options to remain in the area.

Understanding your 'worst case scenario' is also important as well as having just an all around understanding of all the terms you are agreeing to. One area we try (and you should to) to accomplish is to have no surprises when it comes to the partnership tract.

Many Expectations. It is much easier to clarify up front, find out the expectations of the practice and if necessary, cut your losses and move to a better potential future opportunity (a negotiation that doesn't work out still takes your time and/or money to review).

Back to the ENT buy-in…

The physician owner we were working with had never done a buy-in before. After careful questioning, we found out there had been a previous physician employee in the past but it was only a short term position.

The other physician hadn't stuck around to become a partner. This immediately caused us to dig in more and find out what the situation was. Turns out it was a family situation and not an intrinsic issue with the practice on why the physician didn't stay employed longer or become a partner.

As we discussed the contract with the practice owner (we often do so with the owner and not a CEO/COO/Director or attorney with smaller groups) we found quickly how proud he was about the practice and the business he had built. This is often the case with smaller groups.

Sweat Equity. The owner has much sweat equity in the practice and is very proud of the business. As we discussed more, they had 3 locations, 20+ employees, and a nice 7 figure revenue stream – and this physician had created it all from nothing. Obviously, something to be proud of.

Our discussion typically begins with general details of the employment. Benefits, practice set up, etc. We then move to basic contract structure and terms such as the non-compete or 'tail' insurance coverage.

We typically finish with a discussion of compensation and partnership. When discussing the compensation structure with any group or employer, having access to data is very powerful.

What is the average compensation for the area? 75th percentile? 90th percentile? What is the average overhead rate for collections and the average conversion factor per wRVU in addition to the average wRVU production per physician?

Increasing Negotiation Tactics. Understanding what these are in a given geography can greatly increase the chance you are able to negotiation better financial terms with the employer. When we work with smaller groups, they often don't know this information so our detailed discussion is very important. Typically the information found online or in survey data is biased – it takes someone that knows the market and works with many physicians in the area – not just a 'legal' mind.

Once we are finished with the compensation terms and have agreed on what we think is a fair deal for both parties (it can't be too favorable for one party or long-term things just won't work), we move to the partnership discussion.

Knowing there is nothing guaranteed about becoming partner is important. It is very rare to see specific buy-in information in an employee contract so the up-front conversation becomes more important.

Ask Specific Questions. As discussed earlier, there are very specific questions that should be asked on each partnership contract discussion you have.

In our example here, we quickly found out that the physician hadn't given much thought to the purchase price/buy-in amount or how he would even value the practice at the time.

The discussion around what percent he would be willing to give up was even more alarming. While the physician had expected an equal 50/50 partnership the physician was thinking of around 30/70. He wanted to retain majority control!

If you go into a situation expecting 50/50 and come out with a 30% minority stake you would be very disappointed. In essence, throughout the negotiation and discussion we found out that the physician candidate would have a minority stake in the company, have really no formal say in the direction of the company, but would retain a small split of the profitability of the company for a buy-in amount which was undefined. This situation was not acceptable to the physician client. There was just too much that could go wrong.

How It Was Valued. In our discussions we found out he 'may' value the business on a multiple of EBITDA (earnings before interest, taxes, depreciation and amortization) which is in essence the profitability of the company. Or he 'may' value it based on his own personal feelings on the value of the company at the time. Or he 'may' take another route in determining the buy-in price. It was just too fluid.

How It Went Wrong. These numbers can be somewhat manipulated to a buyer that doesn't understand what he or she is looking at. Some businesses sell at a 1 multiple and others at a 10+ multiple (so a business profiting $1MM per year would be worth $10MM). He was going to add in a 'reasonable amount for goodwill' - again, undefined.

How Details Were Sparse. When asked about an equal partnership, he simply stated 'we would maybe be able to get there over multiple years but that buy in would simply be HUGE and he couldn't afford it' – again not enough details for the physician interested in the position.

None of this to say the practice wasn't successful or the owner physician not a great person to work for. This may be a great opportunity for any physician interested. The terms and details of the contract are simply that – details. They may or may not affect your happiness at the group/practice.

However, if you go into a situation expecting one thing and multiple years later you are offered something different, it can be a costly lesson in terms of lost years and lost income (typically these private practices pay less the first few years than a hospital employed position).

The most important thing is you have the right questions to ask and the right amount of understanding to ensure you are making the best available decision with the information in front of you.

Even the best contract and up front questioning can result in a sub-optimal situation but the chances are much reduced. Make sure you enter any partnership discussions armed with the right questions. Assume good intent but think "trust but verify!"

Even though none of the answers may be 'guaranteed' or in the contract it is necessary to have this discussion up front to ensure you are dedicating yourself to the position you think it is.

STORY 2 – 'JUST TRUST ME'

Often people just want to have their contracts looked at for any big red flags as they feel they understand what is included in the agreement. Often times it's just as important what is NOT included in the terms. Sometimes not knowing can be costly.

We recently took a call from a Dermatology MOHS surgeon who was looking to change positions. While he has a new offer from an exciting new partnership opportunity, he didn't have his initial contract reviewed.

What The Issue Was. He informed us he has 'just a 90 day notice I have to give, that's what the contract says.' After discussing the potential issues with not knowing how to terminate his current contract, he allowed us to review the previous one and help formulate a transition plan. What we found during the quick review of his currently active contract was very important.

What Was Left Out. While the contract read well and contained no 'red flags' or major issues with the content included (and he was right – there was a 90 day no-cause termination provision!) there was something that was left out during the initial review/negotiation by him.

The contract provided him a base salary and the ability to earn incentive compensation above a certain threshold. These are common for many specialties, MOHS Surgery being one of them. The problem wasn't necessarily how the compensation was set up, or how much the physician was earning, but how it was structured should he decide to leave the practice.

What The Problem Was with Collections. When there are incentive plans in contracts, it can be important to remember that just because a Physician sees a patient today does not mean the practice is credited with the collections that day. It may be 24 days, it may be 68 days, but (unless a cash-operation like many in Plastic Surgery) it is not the same day or the next.

The Clause. Upon termination, there needs to be a clause that allows you to continue the incentive compensation plan even if you are no longer working at the practice.

If your Account Receivable (AR) balance is $200k and you are to receive 30% of it when it comes in over the next 30-50 days, you'll receive around $60k as it is collected.

The Problem. Many employers state in their contracts that you need to be employed in order to receive the collection bonus. Should this be the case, you'd not receive any of the $60k you had rightfully earned.

There are many employment contracts that state the last day worked is the last day paid. Still others have situations and policies on being employed the day the bonus is paid.

The Difference. Others have clauses based on how and when the agreement ends and how any compensation is paid (some even how expenses need to be paid back to the employer!).

Back to the MOHS Surgeon...

With a large AR balance, he realized that while his lower salary and higher incentive compensation was a good deal so long as he worked there.

It was less than ideal if/when he left. With his particular structure, he was looking at walking away from over $125k in compensation.

Some structures allow you to be paid throughout the term on a monthly basis, some quarterly and some only once per year annually.

There are definitely rights and wrongs with how these important details can be worked out. Make sure you not only understand the compensation structure and all its details, but how you are paid if/when the relationship ends for any reason.

It is much easier to talk through issues on the front end then when more emotions (and money) are on the table at the end of a relationship. Always ask plenty of questions throughout the contract review process.

STORY 3 - THE TAIL HORROR STORY

Picture this, imagine this.

You just started a new position. You are so excited! You HATED the last job with a passion. You couldn't wait to leave.

You're whistling and humming. You are so happy! You go to check the mailbox- see a letter addressed to you. You open it up.

WHAM! It hits you like a freight train. A bill for $184,000.

Your jaw hits the floor. Your employer's tail coverage dropped and now you are responsible to cover it.

Say what??? I thought I was covered in my new gig!!!

One of the biggest financial pitfalls that physicians can get trapped into as they are transitioning into practice or changing jobs can be malpractice insurance.

Biggest Financial Pitfall. The biggest financial pitfall I think in general is not being financially saavy.

It's not knowing... It's not having a plan ... with their financial windfall as far as monthly paychecks and bonuses.

Biggest Goals You Have. Will you pay off debt? Do you buy a new house? Buy a new car?

I think doctors don't have a good idea to plan as far as how to move this process forward as they are now an attending with a much more healthy salary.

For me, from a contract perspective it's the same thing.

Biggest Contract Provisions. How do you make sure that the money that's coming in is the money that you've earned? How do you know that no matter when the contract ends or is terminated that you're going to receive that money?

If your base salary is $200,000, how do you set up a budget for your monthly spend at below 200,000?

We have a physician who was a two year OB doctor in the DC area and was transitioning down to Miami near South Beach and they had a tail out of their policy in DC.

They didn't offer a policy in Miami because they didn't carry malpractice insurance out there for a variety of reasons. So she had to tail out of their tail. She was two years out of training. An OB in DC makes 'ok money', but not tremendous money to put a ton of money in the bank every month. Her tail insurance that she had to pay was $184,000.

Here she is two years out of training. You can imagine that she still has student loans. She is still working her ways towards saving for her first home, starting to save for retirement.

Then, she gets hit for an $184,000 bill? Can you imagine how ticked you would be?

You can't finance tail insurance, right? There's no way to take it back. They can't take it back if you don't make your payment.

You look at this and you say, 'Okay maybe if I've been financially astute, maybe invest a little bit. Put some money in the bank and then maybe have $30,000 liquid, maybe $40,000 liquid'. [Then] you get a bill for $180,000. And that obviously dramatically changes the trajectory of your financial future and the physician didn't even have a plea until they involved us with what their obligations were.

So what was the solution with this gal?

Where she got this big bill, did the employer ended up making up for some of that because they didn't have malpractice?

Did they just try to deal with the previous employer or the insurance company?

What was the solution on that story?

I think if she would have involved us two years prior [it would have helped]. I don't know if we would be able to make the situation go away.

I think we would be able to make that much more palatable. I think we probably could have saved her at least $50,000 or $100,000 if she would have got us involved right away.

Although, having involved us in the back end was better than too late. We were actually involved with the malpractice broker that we consult with. We shopped the tail. We were able to do a negotiation of her way out.

I think we end up saving her maybe $30,000 or $50,000. We helped her out a little bit, but it still wasn't what she need at the end of the day.

She just had to go to friends and family and bank notes to get the money and she just had to pay it. Maybe we could have made it all go away if she would have engaged us in the front end and then she would have a reasonable expectation.

She would be able to financially plan for this event and she wouldn't have been blindsided on a Friday afternoon.

STORY 4 - THE MOVING CARDIOLOGIST HORROR STORY

The Story. There was once a cardiologist that we work with who joined a private practice. He moved half-way across the country in order to land this new job which he was so excited about.

Unfortunately, he did not have the contract reviewed ahead of time. He did not understand it. He didn't come in to negotiate it. He told me later on that he felt so comfortable with the group, it seemed like a natural fit.

He had looked around for several opportunities and he felt this was the best one. It was a private practice clear across the country.

The Error. He flew out there and he agreed to accept their compensation and he was so excited that he signed the contract, without having anyone review it.

He felt it was a great job. The family liked the area. He relocates the family clear across the country too of course.

He gets in the practices, grinds and grinds and works there for about half a year. It was at two-year contract, in which case he was supposedly going to be a partner at that private practice.

The Desire. Here's a guy who wants to be the owner of the business. He wants to make his own hours. He doesn't want to be told what to do by a large public hospital.

He felt this position was a great opportunity for that. Six months into the gig after he finalized the contract, the physicians call him in, sit him down, and they say, "We've sold off all of the practice to a local hospital. And we feel it is the best for the community. We feel it's best for our patients. We feel it is best for our own interests. Unfortunately, they don't need another cardiologist. So, they're just bringing us over. We're going to evoke our "No Cause Verdict."

The Shocking Result. I believe it was 60 days and they terminated him. He didn't have a job.

That's NOT the alarming part. The alarming part was he had the conversation with them about coming on board, never was there mentioned the potential for a hospital integration.

Obviously, the existing partners had been talking about this for quite some time. He didn't have that conversation with what would have been a typical negotiation process of the contract. In addition, because he didn't have the contract reviewed, he didn't fully understand it.

The Ugly Additional Problem. To add insult to injury, not only was he out of a job after 60 days, but, he also had to repay some of his bonuses because he didn't fulfill a year's commitment to them in the practice.

Can you imagine what that would feel like???

He had to re-pay the $10,000.00 that they gave him for relocation. He also had to repay back a $20,000.00 signing bonus in full.

On top of all that, he had a "Non-Compete." Even more infuriating, the "Non-Compete" was all inclusive.

They told him that "we were told we should force the non-compete."

The Move AGAIN. He HAD to move his family AGAIN. He couldn't even stick around the same area because of the non-compete.

Here's a guy who just relocated clear across the country. He took some up-front money, which was gone in an instant. All of this occurred didn't ask the right questions right off the front end.

You can imagine six months into his professional career. His life was dramatically changed.

Here are a few questions that he should have asked before coming on board...

➢ Have you ever had discussions with a hospital, for integrating the practice in joining the group?
➢ Has a hospital ever approached you, with these desires, to employ you, or as a group?
➢ If they have, what did the process look like?
➢ What are your goals long term?
➢ Do you want to keep a private practice in the longer term?
➢ What do the financials of the practice look like? Is the profitability of the practice sustainable?
➢ Do you want to grow into twenty physicians?
➢ Do you want to eventually retire and sell your assets?
➢ Do you want to sell your business to another physician?

There's lots of questions that can be asked around that. When we ask questions, we assume people are going to give us fairly honest and open and candid answer.

At the end of the day, it just depends on what kind of folks you're working with. While we could write an entire book on these 'horror stories' we prefer to look at the brighter side of thins. Just know there should be caution in your frame when looking into any position. Having the right person by your side to understand and discuss the agreement with can be invaluable – seek help and an ear. It's always a good investment.

FINAL THOUGHTS

As a physician, sometimes you may feel more like a circus performer. You are juggling this ball and that ball and that ball. You are literally going from crisis to crisis, performing in a 3 ring circus.

It can be so easy to forget the details like contracts or tail coverage or financial planning.

The huge lesson from these stories: If you are considering changing employers, make sure to ask LOTS of questions, make sure you understand malpractice coverage, make sure you understand all the nons, and make sure you understand what would happen if you left your employer.

You don't want to get hit with a five figure or even six figure bill!

Even more importantly, make sure you are working with professionals who can help guide you through this maze so that you don't become the next horror story.

Learning the hard way is not the way to educate yourself on what a bad situation/contract looks like.

The Wrap-Up

CHAPTER 33

SECRETS TO FINANCIAL SUCCESS FOR DOCS 2.0

By Dave Denniston, CFA

H ave you ever looked back at something you did and thought.... "Hmmm... I bet I could do this better!"

I got to this point where I've written a ton of chapters already for this book and had some editing done. Yet, it felt like something is missing...

<u>What could I be missing???</u>

It was nagging at me like an itch that I can't scratch. It was driving me nuts!

Then, I had a thought.... Maybe I can tie this altogether.

In the book, we talk about financial mindset, paying off your debts, buying your first home, negotiating your contract, and Investments 101.

But….

We don't give any guidance on how this should get done.

<u>After all, how can a busy, busy physician balance all these priorities?</u>

Furthermore…

- *What's the formula for success that physicians could follow?*

- *What are some case studies that could show the best places money should be allocated to?*

So, I dusted off the first article I EVER wrote. It's called "Secrets to Financial Success for Young Physicians."

I thought it was a great start and in this chapter I'm going to expand on it more and more.

We're tackling the questions…

- *What should your cash flow look like?*

- *How does a physician budget for one that DOES have student loans compare to those that DON'T have any student loans?*

- *How does a specialist compare to a primary care physician?*

Anyhow, with a bit of dusting off and no further adieu, here is the all new Version 2.0….

. . . .

THE OVERVIEW OF SECRETS TO SUCCESS 2.0

Imagine you have just finished residency and are finally starting your medical career. You may feel relieved to finally have those years of schooling and training behind you. But then reality sets in and you wonder how you are going to deal with the medical school debt that you've accrued.

According to the American Medical School Association[1], 86% of medical school graduate carry educational debt and that the median debt burden is $155,000 for public school graduates and nearly $185,000 for private school graduates.

How can a young physician balance the financial stresses of overwhelming medical school debt with buying their first home, saving for retirement, buying a new car, and at the same time enjoying life?

Let's approach this with a specific plan of action with some principles that can help any young physician. I have broken down the prioritization based upon multiple case studies- several for primary care physicians and several for specialty physicians.

We all know that specialist physicians on average, have more income than a primary care physician. Due to the extra wiggle room, there are some significant differences between the two types of case studies. For example, we bump up retirement savings and a down payment for the purchase of a first home.

To develop a plan, you will need to take some time to ponder and reflect upon your situation.

<u>The best time to do that is now, early in your career.</u>

Here are some basic guidelines that I want you to consider IN this specific order.

Again, this is THE Formula that you want to follow in this SPECIFIC order...

FOR RESIDENTS

- ➢ Step One: Save In A Cash Cushion/Rainy Day Fund
- ➢ Step Two: Grab Your Free Money
- ➢ Step Three: Pay Down Highest Priority Consumer Debt (Including Student Debt)
- ➢ Step Four: Save More in Cash Cushion/Rainy Day to fund short-term goals

FOR SUPER-STAR RESIDENTS/FELLOWS & PHYSICIANS TRANSITIONED TO PRACTICE

- ➢ Step Five: Put More in the 401k/403b
- ➢ Step Six: Fund the Roth IRA or Back-Door Roth IRA
- ➢ Step Seven: Consider Starting a Side Hustle
- ➢ Step Eight: Pay Off All Consumer/Student Debt
- ➢ Step Nine: Sock Away More Non-Qual $
- ➢ Step Ten: Consider Alternative Investments- Peer to Peer Lending/ Real Estate/ etc
- ➢ Step Elven: Max out 401k (consider Roth 401K INSTEAD) and Max out 457 DC (if available)
- ➢ Step Twelve: Sock Away More Non-Qual $
- ➢ Step Thirteen: Pay Off Your Mortgage
- ➢ Step Fourteen: Dang! You're Awesome! You might need more tax-deferral strategies. Come see me for more. They may or may not be a good fit for your situation

Case Study# 1	
Single Resident Physician with $250k Student Debt/ IBR	
Income	**$ 55,000**
Taxes[3]	$ 11,168
Rent	$ 9,000
Living Expenses	$ 24,000
Student Debt	$ 5,580
Rainy Day Fund	$ 3,602
Save for First Home	$ n/a
Retirement/ 401k	$ 1,650
Expenses	**$ 55,000**

Case Study# 2	
2 Married Resident Physician with $500k combined Student Debt/ PAYE	
Income	**$ 110,000**
Taxes[3]	$ 22,336
Rent	$ 18,000
Living Expenses	$ 48,000
Student Debt	$ 7,440
Rainy Day Fund	$ 10,924
Save for First Home	$ n/a
Retirement/ 401k	$ 3,300
Expenses	**$ 110,000**

So, let's break this down a bit…

If you are single, I am assuming that you are splitting the rent with a friend. Let's say a place would cost $1,500/month on your own. Your portion of that cost is split in ½ with a friend is $750/month in case study# 1.

It is duly noted that in case study# 1 and #2, we are barely making ends meet. We are purely focusing on steps# 1 through #4. There's not a whole lot of wiggle room.

In my opinion, one of the most important things you need to do is to lower and eliminate their consumer and educational debts as well as to establish your "rainy day" fund.

Now that you have your first job as a physician, make sure for the first few months to set aside money for the "stuff happens" factors in life like the car breaking down or the furnace going out. I call this the "rainy day" fund.

Keep these funds in a money market or checking account until you have more than $15,000. Then, consider a low-risk investment account where you can pull out ALL of your money without any penalty, if necessary.

ENJOYING A REASONABLE LIFESTYLE

What is a reasonable lifestyle? Although this means different things to different people, I consider it to mean that you aren't just squeaking by.

As a resident, to me, this means that you are able to eat out from time to time, go on a local stay-cation, or do things you enjoy while still being committed to keeping your expenses within limits.

To help keep an eye on your living expenses, use free budgeting and wealth management tools such as Mint.com and creditkarma.com. (We use emoney advisor at my office)

WHAT IS GRAB THAT FREE MONEY??

Step# 2 is to 'Grab That Free Money!". You may be asking, what's that all about?

In your employer sponsored retirement plan- like a 401k or 403b- many residency programs give you free money.

Say what????

Yes, many (but not all) will give you free money for participating in the retirement plan.

Make sure to contribute at least up to the maximum match that your employer provides (if any).

If your employer matches dollar for dollar, this is like an automatic 100% return. Even if your employer matches 50 cents or 25 cents on the dollar, that is still like a 50% or 25% return just for contributing.

Grab that free money my friends!

Hint: As a resident, the money you put into the traditional 401k/403b is tax deductible. This can be helpful for minimizing payments to your debt forgiveness programs.

On the other hand, if you don't have any student debt OR aren't going for debt forgiveness, you may be better off contributing to the Roth 401k/403b instead since you are likely in a relatively low tax bracket.

WHAT ABOUT SAVING FOR MY FIRST HOUSE?

At this stage, we're not likely able to save for a house, but we are (hopefully) socking away some for the rainy day fund that we have some money to fall back on for emergencies.

You add enough to that rainy day fund and now we're thinking of the first house! I generally suggest $10,000 to $15,000 as a substantial cash cushion.

This could allow you to buy a car outright with cash or fix this or that emergency that pops up.

If you don't have ANY student debt, then you can certainly move down to steps# 5 and 6 and save more in the 401k or Roth IRA or purely focus on adding to the cash cushion/ buy first home fund.

NOTE: As a resident when you contribute outright to a Roth (assuming you meet the income limits- which 95% of residents will), the first $10,000 of principal AND earnings can be withdrawn for a purchase of your first home.

Otherwise, for other withdrawals- like a car or simply for living expenses your earnings could be subject to penalties.

Case Study# 3	
Single Practicing Primary Care Physician with $250k- Student Debt/ IBR	
Income	**$ 170,000**
Taxes[3]	$ 49,590
Rent	$ 18,000
Living Expenses	$ 49,486
Student Debt- IBR	$ 22,824
Rainy Day	$ 6,000
401k- Match $	$ 5,100
Save for First Home	$ 14,000
Extra 401k	$ 5,000
Expenses	**$ 170,000**

Case Study# 4	
Married Specialist Physician (Sole Income) with $300k- Student Debt- PAYE	
Income	**$ 300,000**
Taxes[3]	$ 71,801
Rent/ Mortgage	$ 24,000
Living Expenses	$ 75,000
Student Debt	$ 27,600
Rainy Day/ Further Debt Paydown	$ 12,000
Save for First Home	$ 30,000
401k- Match $	$ 9,000
401k- Extra	$ 9,500
Back Door Roth	$ 11,000
Debt Pay-Off	$ 30,099
Expenses	**$ 300,000**

In case studies# 3 and #4, we explore two situations where we have a sole income.

Now these physicians have each transitioned to practice and we start to be able to move up the ladder of success as we climb up the steps.

I make a few dangerous assumptions here.

First, I assume that your rent is 'reasonable' at $1,500/month to $2,000/month. Certainly, it cities like New York & San Francisco this can be a drop in the bucket.

Next, I throw a bunch of costs into living expenses like term life insurance, disability income, and so on. Thus, living expenses aren't simply groceries and vacations. Make sure to keep that in mind that we've separated rent/mortgage from living expenses. If we add together rent/living expenses, we're looking at about $6,000/month for case study # 3 and $8,000/month in case study# 4.

However, the good news here is that in case study# 3 this physician has saved about $30,000 out of $170,000 income or close to 16%. This isn't a bad savings rate and this physician will be on the right track.

In case study# 4, we've bumped up living expenses to reflect an extra person in the household. Since this person is a specialist, they are able to move up the steps of success even more.

They are to utilize some additional really cool tactics like the back door Roth & make a huge dent in their student debt & save a ton for a house.

Including the additional (i.e. optional) debt pay-down, they are saving about $100k out of $400k or nearly 25%! This physician is on the fast track to success. I am super stoked for them!

Case Study# 5	
Married Physicians (Dual Income) with a Kid with $500k- Student Debt- PAYE	
Income	**$ 400,000**
Taxes[3]	$ 99,180
Rent/ Mortgage	$ 24,000
Living Expenses	$ 75,000
Nanny	$ 30,000
Student Debt	$ 45,648
Rainy Day	$ 12,000
Save for First Home	$ 30,000
401k- Match $	$ 18,000
401k- Extra	$ 19,000
Back Door Roth	$ 11,000
Debt Pay-Off	$ 36,172
Expenses	**$ 400,000**

Case Study# 6	
Married Physicians (Dual Income) with a Kid with $500k- Student Debt- PAYE	
Income	**$ 400,000**
Taxes[3]	$ 99,180
Mortgage	$ 72,000
Living Expenses	$ 100,000
Nanny	$ 30,000
Student Debt	$ 45,648
Rainy Day	$ 12,000
Save for First Home	$ 23,172
401k- Match $	$ 18,000
401k- Extra	
Back Door Roth	
Debt Pay-Off	
Expenses	**$ 400,000**

We've talked a ton in this book about three words: LOCATION, LOCATION, and LOCATION.

My friends, where you live can determine when you can retire.

Let's drive this point home again in case studies# 5 & #6.

In case study# 5, we have a two physician household. Let's say that they are two primary care physicians. They are making great money and have a kiddo or two.

They live in the Midwest where rents/homes are easily affordable in their budget.

As a matter of fact, they bought a comfortable spacious 3,000 square foot home for $300,000 that makes their good physician friends who live on the eastern seaboard incredibly jealous.

However, they have a nanny to help manage their family with their incredible busy, compressed schedule.

They have a ton of debt and between the two of them are making payments of nearly $4,000/month.

When we add up the living expenses and rent/mortgage and nanny costs, they are spending about $11,000/month. It's an expensive household, but they are making it work and enjoying life.

On top of that, they are also able to climb the steps to success. They are not only saving in their 401ks, they are maxing them out! On top of that, they are saving the max in the their Back Door Roth IRAs.

Not only that, they are knocking out their debt beyond the monthly payments.

If we add up all of their savings and optional debt repayments, they are socking away about $115,000 a year or nearly 28%.

They are on track to achieve financial freedom in their early 50's and they are super stoked and excited!

Now, let's compare this couple to the married physicians in case study# 6.

These physicians are amazing doctors. They love their patients. They love where they live.

They are living in beautiful San Diego with beaches a plenty, fantastic foodie joints, and places filled with loads of activities for the kids.

It's an amazing place to live!

In addition, they have deep roots there. There's plenty of family support and loads of friends.

Moving away was tempting, but it was too hard to leave that support system they built up over the years.

Unfortunately, their extremely expensive, but modest home cost them $800,000. The downpayment barely made a dent in the debt that they accumulated in buying the place. The mortgage payments are killer, but they think it's worth the trade-off.

Consider that this couple saved, which is awesome that they did practice the habit of saving, a total of $53,000.

However, living in an amazing place comes at a cost. Food costs more. The entertainment costs more. The property taxes cost more.

At a mere 13% savings of their income (which in the big schemes of things is still OK), this couple is going to have to work well into their 60's and will be relying on selling their home to achieve a comfortable retirement away from San Diego.

They aren't able to afford saving in back door Roth IRAs and they can't max out their 401ks.

There's a tremendous difference!

Case Study# 7	
Married Specialist Physician (Sole Income) with NO Student Debt	
Income	**$ 400,000**
Taxes[3]	$99,180
Mortgage	$30,000
Living Expenses	$75,000
Student Debt	
Rainy Day	$12,000
Save for First Home	$60,000

Case Study# 7	
401k- Match $	$12,000
401k- Extra	$6,000
Back Door Roth	$11,000
NQ Savings	$70,000
457 DC	$18,000
Alternative Invest	$6,820
Expenses	**$ 400,000**

Case Study# 8	
Single Practicing Primary Care Physician with NO Student Debt	
Income	**$ 170,000**
Taxes[3]	$ 49,590
Rent	$ 24,000
Living Expenses	$ 55,000
Student Debt- IBR	$ 0
Rainy Day	$ 12,000
401k- Match $	$ 5,100
Save for First Home	$ 14,000
Extra 401k	$ 5,000
Additional NQ Savings	$ 5,310
Expenses	**$ 170,000**

In our last two case studies, case study# 7 and 8, we look at how to allocate money when a physician has NO debt.

Certainly, the specialist physician has a HUGE advantage here with the extra dollars they are earning.

However, the primary care physician is still doing amazingly well. They aren't going to be able to retire in the next 5 to 10 years, but they are able to be laser like focused on their goals.

These are a ton of case studies my friends! Take some time to look over what you are spending and how it stacks up to our case studies.

As we wrap up this book and this chapter, let me give you a few thoughts to ponder about spending and spending habits.

BE SMART ABOUT BIG-TICKET ITEMS

After years of living in small houses or apartments or driving older vehicles, many young physicians feel they're more than ready to purchase a new home or car when they finally get their first "real" job.

Instead of rushing out to buy such big-ticket items, take a moment and think about how those purchases might affect your ability to achieve your long-term financial goals. And bear in mind that your lifestyle can change as you pay off your debt.

Here are a few of things to consider when you are ready to make those purchases:

The Trap. When buying a home, make sure you can put at least 20 percent down on the property. If you put only 5 or 10 percent down, you may be required to have private mortgage insurance (PMI), the cost of which could raise your monthly payment by a couple of hundred dollars.

Note that there are some physician specific loan programs that may not require PMI. However, getting in the practice of savings towards a goal is a wonderful form of financial discipline.

The Money Pitfall. Remember that cars are depreciating assets. The second you drive a new car off the lot, you typically lose $5,000 to $10,000 of the value. Why put a significant chunk of your hard earned dough in something that you know you will lose money on?

Instead of buying a new car, consider buying a used one and holding onto it as long as possible. Financially speaking, buying a low-mileage used car (say with 20,000 to 50,000 miles on the odometer) and holding it for five years or more makes much more sense than leasing or buying new.

Use Cash. I strongly suggest paying cash for a car. If you already have a high-interest car loan, consider paying it off as soon as possible. If you do decide to get a new car, remember that buying can be a better deal than leasing, especially if you hold on to the car for five years or more. If you lease a $20,000 car over three years at 6 percent interest and pay $1,000 down, the total cost over three years will be $12,600 plus the down payment.

Why Leasing Is Inferior. At the end of the lease, you will have paid $4,200 toward the principal of the loan and can either purchase the car or return it to the dealer. If you purchased the same $20,000 vehicle with the same down payment and finance it at 6% interest, you would pay $7,500 per year ($22,500, plus the down payment over three years). At the end of the loan period, you will own the car.#2

If you look at the cost of leasing over 10 years (let's say that you renew your lease every three years) and get the latest model car, the costs for leasing will be at least $42,000 plus down payments. Whereas, if you bought the car and held on to it, the cost will have been $22,500 plus regular maintenance. Buying rather than leasing would save you nearly $20,000.

FINAL THOUGHTS

Physicians have big hearts and big dreams. They may want to pay for their kids' college education, buy the cabin or second home, buy a boat or RV, or give lots of money to worthwhile charities.

My advice: Hold off on these things until you are debt-free. Remember that once your debts are paid, you'll have the cash to fund these other projects.

As a physician, you've made a commitment to helping others and your community.

Now you need to make a similar commitment to your finances.

If, as a young physician, you focus on paying off your debts, save for a rainy day, live within your means and put money away for retirement, you can then do the things you've long dreamed of doing and be well down the road to financial independence.

I would love to hear from you, comment or e-mail me at <u>dave@ doctorfreedompodcast.com</u>

CHAPTER 34

THE FINAL WRAP UP
By Dave Denniston, CFA

As I write this chapter in late November 2016, I recently received news that our friend and co-author Amanda Liu passed away.

I am shocked and devastated and heart broken.

I struggled with the thoughts of what to include in this book. There were several chapters in the original manuscript that I've been wondering about. They hit close to home as she revealed her struggles in this text.

I was considering… Should we take out some of her chapters? All of her chapters? How can we honor her?

She was raw and real. To take away that aspect would not serve who she was.

I think the best way to honor Amanda was to continue her mission of educating and inspiring doctors.

Amanda noted on her blog…

"When I count my blessing, gratitude replaces fear.

Gratitude is key to MY ability to enjoy life at the present instead of worrying about the future.

'Depression is living in the past; anxiety is living in the future; peace is living NOW.' *said by an unnamed schizophrenic patient I met on my psych rotation."*

My friends, money is a tool. It is meant to help us enjoy life and live to the fullest. Let us have gratitude and express thanks for all that we have.

In this book, we've laid out a roadmap as best we could to helping you achieve your potential.

I encourage you to share the message and continue the circle of giving.

I am inspired by the legacy of her words and her mission.

Let us know how this book has impacted you and how Amanda's words have impacted you.

You can contact me at <u>dave@doctorfreedompodcast.com</u>